Modern
Horse
Breeding

This book is dedicated to my elegant,
talented and quirky
Thoroughbred mare
SARAH
(GSB Curly Wee)
for everything she was,
everything she gave me,
and for putting me back on track.
Her genes live on.

Modern Horse Breeding

A GUIDE FOR OWNERS

Susan McBane

Veterinary Consultant: Janet L. Eley, BVSc, MRCVS
Stud Management Consultant: Barrie Hosie
Stud Groom to The Duke of Roxburgh

SWAN·HILL
PRESS

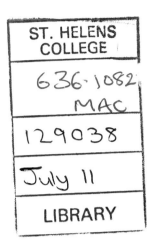
Copyright © 2000 Susan McBane
First published in the UK in 2000

by Swan Hill Press, an imprint of Airlife Publishing Ltd

British Library Cataloguing-in-Publication Data
A catalogue record for this book
is available from the British Library

ISBN 1 84037 032 7

Typeset by Servis Filmsetting Ltd, Manchester, England
Printed in England by Butler & Tanner Ltd, Frome and London.

Swan Hill Press
an imprint of Airlife Publishing Ltd
101 Longden Road, Shrewsbury, SY3 9EB, England
E-mail: airlife@airlifebooks.com
Website: www.airlifebooks.com

ACKNOWLEDGEMENTS

There are two people without whom this book would not have the ring of authority which it has, and whom I should like to thank most sincerely for their conscientious and patient attention to my queries and for checking their sections of the manuscript. The first is Janet Eley, BVSc, MRCVS, in my view one of the best equine veterinary surgeons in the UK, for her unfailing help and advice with Part 1. The second is Barrie Hosie, one of the few people to whom I would entrust my horses and who has, for many years, been Stud Groom to the Duke of Roxburgh, for similarly helping with Part 2. Any remaining errors or omissions in the book must be mine.

A third person deserves my gratitude, and that is Captain Elwyn Hartley Edwards, without whom I should not have had the opportunity to write this book in the first place, and I also owe considerable thanks to the editorial team at Airlife Publishing for their professionalism and patience during the entire process.

Dianne Breeze's drawings, as always, add a most attractive artistic clarity to what can be a tricky subject to put over in words.

Vanessa Britton supplied many of the photographs at short notice, for which I thank her.

Finally I wish to thank Hilary Self of Hilton Herbs (see Appendix) for her help and advice on herbal products and complementary therapy practice and her time in checking the appropriate part of the manuscript.

CONTENTS

Irish Mist.
Dawn on a Thoroughbred stud in Ireland. (Tim Hannan)

PART 1
VETERINARY CONSIDERATIONS

Chapter 1
Why Do You Need a Vet?

Most people involved in breeding horses regard the whole process as one of nature's miracles. It is so stunning, convoluted and complex that it could not possibly have been dreamed up by a human being even for the most fantastic science fiction novel.

From the very beginning, when mare and stallion first perceive the attractive, stimulating pheromones or scents each gives off during the mating season, especially the mare when she is in oestrus (also called 'in heat', 'in season' or, an older term, 'horsing'), to the end of the process which can be said to be when the foal is finally weaned from its dam, it is just one incredible occurrence after another.

From the joining of two individual, microscopic cells – the ovum or egg from the mare and the even smaller spermatozoon or sperm cell from the stallion – develops, in a world of its own (the sealed environment of the mare's reproductive tract) an embryo, which develops into a foetus, which finally becomes the foal we hope will grow up into anything from a much-loved family pet to an Olympic or Derby winner.

The process of cell division and multiplication, which causes the original two cells to become countless billions of other specialised cells forming different parts of the body with different purposes, progresses according to the genetic code or 'instruction leaflet' which ensures that the foal will first be a foal and not a puppy or human baby, for instance, and secondly take after its parents or other ancestors, according to which genes it receives. Of course, reproduction is a perfectly natural process and although, even in natural conditions, mating by no means always results in a pregnancy, by the time the process reaches the foaling stage 90 per cent of the time it goes off without a hitch. Novice breeders, for instance, try as they might to oversee the birth (not only in case something goes wrong but also so that they can actually witness this miracle), are often fooled by the mare, whose instincts tell her she needs to be alone, and very often they miss this great event. Sometimes, when sitting up with her, they pop off to make a cup of coffee, certain that she is not going to foal imminently, only to find the foal on the floor already trying to get up when they return. 'You weren't needed,' the mare seems to say. 'We've managed fine, thank you.'

Things can, however, occasionally go wrong in a short space of time. For example, the foal can be presented in the wrong position, it can be strangled by the umbilical cord, it can be deprived of oxygen during the birth process or it can seriously injure the mare's reproductive tract. In some situations where things do not go smoothly, just a very few minutes can mean the difference between life and death for the mare and/or the foal. Not only do you need expert help at such a time, most probably from your veterinary surgeon, you also need to understand the signs that indicate whether all is right or not, so that you can call for help in good time.

On larger studs, knowledgeable, experienced staff can act effectively in many cases involving minor problems, but they are probably quicker to call the vet when there is a situation they cannot handle than owners foaling down mares themselves. They probably appreciate all too well, from past experience, just how crucial veterinary knowledge and assistance can be. It is easy to decide to 'wait and see' and hope the mare will get herself out of trouble, but apart from being a false economy, this can often cause extreme pain and suffering to both mare and foal, and might result in the death of one or both.

The role of the vet

It is true that veterinary services can seem very expensive to many people these days, particularly those with few horses or just a single mare, but it is something breeders must budget for. Horses are expensive to keep; they are even more expensive to breed, and if the potential cost appears to be too much, you would be better not breeding at all. Not only are vets best placed to keep you up to date with the latest thinking and developments in equine reproduction and stud management, but also they are invaluable sources of advice and information on preventive management as well as on actual disease.

Before you even choose a stallion to send your mare to, your vet can examine her and tell you whether or not she is good broodmare material. Is her conformation suitable for conceiving, carrying to full term and delivering a foal successfully? How is her general health? Does she have any hereditary or other diseases which would physically or ethically preclude her from breeding? Does her past breeding record (if any) indicate that she is going to present you with problems rather than with a healthy, promising foal? These and other questions should all be sorted out before you select a suitable stallion and, if the risks outweigh the potential benefits, any idea of mating her should probably be abandoned.

If you do go ahead with your plans, the vet can help you get your mare into breeding condition and cycling properly and regularly if this does not seem to be happening naturally, particularly if she belongs to a breed such as the Thoroughbred, whose breed society regards 1 January (in the northern hemisphere) as the official birthday of that breed's foals. To the horses, this is midwinter, and breeding could not be further from their minds! Once your mare is cycling, your vet, or the vet used by the stud if you are sending the mare away, can tell you or the stud staff just when she is likely to ovulate and will therefore be ready to mate with the best chance of conceiving, thus saving the stallion's and the staff's resources.

A vet can also confirm whether or not your mare has conceived, and whether or not she stays in foal over the forthcoming weeks and months, tell you how she and the pregnancy are progressing, detect twins (always an undesirable situation) or an abortion you may miss and, as foaling time arrives, give you vital information as to when you can reasonably expect the big event to occur.

As I have said, veterinary intervention can be crucial during the actual foaling process, but even afterwards, his or her role is far from over!

Veterinary expertise will be needed to assess the health and wellbeing of mare and foal.

The most stressful thing that happens to most foals is simply being born. Various conditions can occur at birth, or shortly afterwards, for which you will need a vet's help and advice. After being born, foals have to make massive adaptations, mentally and physically, to living in a completely different environment from the safe, warm, constant one experienced in the mare's womb. They must largely shift for themselves in a cold, harsh world, no matter how good a broodmare their dam is, and they are prey to all sorts of diseases, injuries and disorders which just seem to come from nowhere, as well as to the vagaries of the weather, and the attentions (sometimes unfriendly and sometimes downright injurious) of other animals (not only other horses). They may also have to cope with incorrect management from humans (such as inappropriate feeding or exercise regimes), as well as infestation from parasites against which they have no resistance – in fact, the general stress of living and surviving in the outside world.

As the foal grows, developmental problems can occur and you will need veterinary assistance to cope with these, and probably that of a good farrier if they involve the feet and legs. The mare may be put in foal again and need a constant, watchful eye kept on her and, even after weaning, the foal is not out of the woods. It is still open to the diseases of the young as well as those situations of disease or injury to which any horse is prone.

So even though mating and foaling may be the most natural things in the world, the services of a good, interested equine veterinary surgeon are vital to your own peace of mind and the continuing wellbeing of your mares and foals.

Advances in veterinary science relevant to horse breeding

Apart from increasing knowledge of all equine disease and injury situations, the advances which have taken place over the last few years in our knowledge of matters relating to horse breeding, are considerable and significant – and they continue. Owners of small studs or just one mare may feel that these advances do not apply to them, but in practice they may well do so. You only need one mare to use and make a success of artificial insemination, embryo transfer or DNA testing, for example, and your mare could contribute to the future of her breed or type of horse or to the Equine Genome Project which, like its human equivalent, aims to map and identify equine genes for purposes not only of identification but also of the transmission of good or bad characteristics and disease, all of which will eventually have far-reaching effects on the production and use of horses as a whole.

DNA testing or 'fingerprinting'

The initials DNA stand for deoxyribonucleic acid. DNA is the basic unit of your horse's chromosome construction or genetic 'instruction leaflet', which decrees what characteristics it will have. It is found in the nucleus or 'control centre' of every cell in the body and carries coded instructions from which new cells are made. It consists of two strands or helixes linked together by pairs of sugar and phosphate molecules bonded by hydrogen. These are known as nucleotide bases and they occur in specific sequences.

We can visualise chromosomes as a coiled rope-ladder of DNA with the bases forming the rungs, each rung having a join or bond in the middle. The 'ladder' is divided into segments thousands of bases long, which are individual genes, each coded for (carrying instructions for the creation of) a particular protein which will carry out those instructions.

The individual way in which a horse's DNA is structured can identify it with a percentage accuracy in the high nineties and confirm whether or not its purported parents are likely to have produced it. It is far more accurate than blood-typing and will also be able to be used more and more to detect characteristics both good and bad, and also diseases. In this way, we can determine a horse's physical characteristics, its temperament and whether or not it will suffer from certain diseases or pass them on to its offspring as a 'carrier', without being affected itself.

Genetic screening tests for inherited diseases could become the norm, as well as for identifying 'traits of interest', as they have been described – in other words, a horse's likely abilities in the form of jumping, speed, endurance, action and so on which will make breeding horses safer and more certain. As most breeders know, not all good or bad points are passed on to a foal by its parents, and removing the guesswork in this way when arranging matings will enable you to dispense with the 'wait and see' policy which is prevalent today. The breeding of competition warmbloods and, to some extent, Thoroughbred, Quarter Horse and Standardbred racehorses has been refined considerably by breeders selecting known 'dominant' animals where certain features are concerned, but DNA screening tests will be a virtually foolproof system.

With the increasing importance of registered breeding stock and documentation to a horse being regarded as economically valuable or even saleable, DNA testing or 'fingerprinting' as it is also called, will probably become standard practice when registering horses. The technique probably first hit the world at large during the intensive coverage of the O.J. Simpson murder trial in the USA the 1990s, which was televised around the world. It has been refined into a quite simple test, necessitating only a few pulled hairs, their roots being sent to a laboratory for testing and being required only once in the horse's lifetime. The result is stored on a central computer and the horse's owner receives a certificate bearing a unique code. The UK company, North Western Laboratories in Lancashire, has been given the exclusive European licence to store the information. At the time of writing (1998) the test costs £50.

Artificial Insemination (AI)

Although it is not a new technique by any means, the use of AI is becoming more widespread. It is still not permitted in the Thoroughbred industry, however, owing to the Jockey Club's refusal to accept it for economical and political reasons, reasons which include:

- Because a single ejaculate of a stallion can be used to inseminate several mares (for years after his death if it is frozen), the rarity of his genes will be reduced and so, therefore, will the cost of a 'service' to him, which has economic implications.
- If it is used unwisely in an uncontrolled way, the breed could become overloaded with the genes of just a few popular stallions.
- If more mares are inseminated at home, visiting mares at public studs will become a rarity and this will cause unemployment in the bloodstock industry as the maintenance of large, well-staffed studs will become unnecessary.

Farmers, however, have for generations recognised the advantages of artificial insemination in cattle and pigs and many non-Thoroughbred breeds of horse now use the technique, probably the highest-profile being competition warmbloods of various breeds. The advantages of the method are:

- The risk of injury to valuable mares and stallions during mating is eradicated as the two never get together.
- There is no chance of sexually transmitted disease infection of the stallion by the mare if she has not been correctly swabbed and screened, and little chance of infecting the mare if the semen is properly tested.
- Animals which cannot mate conventionally for some reason, such as aged mares who cannot bear the weight of a stallion, or stallions whose hind legs or backs are failing or who cannot mount or serve a mare properly for some other reason, can still be used for reproduction.
- Frozen semen means that a stallion's ejaculate can be made available to mares all over the world, so obviating the need to transport mares, possibly in foal, long distances for mating.
- If, for performance reasons, a top-class competition horse is to be castrated, his ejaculate can first be stored and his genes used for future insemination, whereas they would otherwise be lost. Moreover, if colts coming into the competition world are to be castrated before being broken in, their sperm can first be taken and stored so that, should they prove to be top competition horses, they can 'sire' foals even though they are geldings.
- Competition stallions can also be used for semen collection for AI, which means they do not need to be retired, usually irrevocably, to stud.
- Depending on the individual horse's temperament, performance stallions do not often combine competing with breeding but may be able to be used for AI whilst still continuing to compete.
- Rare and endangered breeds can be helped to survive by storing stallions' semen for many years to come.
- Mares can stay at home and be inseminated, which is often preferable for the owner who may not wish to be without a much-loved or highly prized mare. This also avoids the trouble and risk of taking a mare (possibly in foal) on a journey, which may be long, and the expense of keep charges.
- The mare can often be inseminated at home at precisely the best time during her cycle for fertilisation to occur, without the need to consider the stallion's other commitments on that day.

The objections to the system, apart from those given previously, come mainly from those horsemasters who prefer horses to have as natural a life as possible and who do not wish to deny them the process of mating. It does smack somewhat of 'factory farming'. There is also the odd objection from mare owners who feel they cannot be sure their mare will receive the correct stallion's ejaculate, but this sort of mistake is very rare in a strictly and ethically controlled system.

Embryo Transfer (ET)

This procedure comes into the same category as artificial insemination as far as breeding Thoroughbreds is concerned: it is not allowed for the same reasons that AI is prohibited. With both AI and ET, though, there is nothing to stop a Thoroughbred animal being used for either or both techniques, provided the stock is not required to be registered in the General Stud Book (GSB) to which all Thoroughbreds in the world must be able to be traced. Many warmblood breeds now contain a great deal of Thoroughbred blood and have pure Thoroughbred stallions (and, to a lesser extent, mares) registered in their stud books, but their stock from warmblood mates (not being pure Thoroughbred, anyway) is registered in the warmblood breed's registry. And two Thoroughbreds which reproduced by means of AI or ET would have the foal registered as a warmblood: it would not be eligible for registry in the GSB.

Owners of non-Thoroughbred stock, however, can reap all the benefits of this modern technology in making breeding much more convenient. The horses are still denied the natural aspect of actually mating, however, which is a disadvantage to some owners.

The general advantages of ET are:

- A mare can 'have a foal' whilst not even being pregnant and whilst still working or competing, because she does not carry her own foal – it is flushed out of her uterus and carried by a surrogate mare.
- Mares who can conceive but are unable to carry a foal to term can still pass on their genes.
- A mare, in principle, can donate several eggs per breeding season without interrupting her competitive career (although it is possible that some registry authorities will only register one of them, to avoid overloading the gene pool with one animal's genes, as with AI). This advantage is, in fact, the main one with embryo transfer. Even if a mare is no longer working, the technique still enables her to provide the embryos

All foals are appealing but you need the services of a good veterinary surgeon to help you maintain them in good health throughout their lives.
(Vanessa Britton)

for several foals per season rather than the usual one per year or two.

The future

Although horses are not considered economically important on a world scale, there is no doubt that, to those involved, they represent considerable outlay. The racing industries, whether involving Thoroughbreds, Standardbreds and other pacers and trotters, or Quarter Horses, are multi-billion-pound industries employing many thousands of people world-wide. The competition horse industry is certainly burgeoning world-wide and top competition stallions can now fetch very high prices – higher than many Thoroughbred racehorse stallions.

Because UK governments, of whatever persuasion, are still unwilling to accept the horse industry (Thoroughbred or otherwise) as being of significant economic importance, despite evidence to the contrary, government funding for research is not forthcoming. There are, however, some very wealthy people and large companies involved in the racing and competition worlds, and what research is done here is financed, directly or indirectly, mainly by such individuals or companies. Research financed by betting taxes and grants from the Horserace Betting Levy Board goes mainly to benefit racehorses and bloodstock but there is a significant advantages in that the knowledge filters through, fairly quickly, to other sectors of the horse world.

Fortunately, research is world-wide and research scientists, although working in their individual countries, regard science as ethically without geographical boundaries, and are not averse to sharing their knowledge although politics sometimes become involved.

Whatever new knowledge, practices, drugs and techniques arise in the future, your veterinary surgeon is the best person to keep you up to date with its implications. Equestrian magazines these days are much more on the ball when it comes to publicising and explaining scientific research quickly, but its application to your own horses should be the province of your veterinary surgeon.

Complementary (sometimes called alternative, natural and holistic) management practices and therapies are also much in favour now, but remember that, in most cases, by law their application needs the initial referral of a vet. Vets are becoming more and more interested in these therapies and it is my experience that few will deny a referral to a likely-sounding therapist. Some vets are qualified not only in veterinary surgery but also as practitioners in certain complementary therapies, often in more than one, although they are still very few in number.

The whole field of treatment of horses, whether by mainstream or complementary medicine, is becoming more open and co-operative, but your starting point whenever you have queries or problems is and should be your veterinary surgeon. He or she can certainly make the difference between success and failure in any breeding enterprise, large or small, and can ensure the ultimate health and wellbeing of your horses and give you peace of mind in what is, without doubt, an undertaking fraught with risk and expense.

Chapter 2

Reproductive Function in the Mare

It is often said that a mare is 60 per cent or more responsible for the way a foal will turn out. Because foals inherit half their genes from their dams and half from their sires, this may sound rather strange but in practice it works out that way.

The condition and health of the mare's reproductive organs can be crucial to whether or not she can conceive, carry a foal to full term, deliver and raise it. The foal obviously develops inside the mare and any problems with her reproductive organs, such as malformation, malfunction or disease, can affect its growth and final physical health and conformation. If, for example, there is a problem with the foetus receiving adequate nutrition via the placenta (the membrane surrounding it in the womb which carries nutrients and oxygen to the foetus and removes its waste products), it will not develop properly. Some infectious diseases suffered by the mare can adversely affect the foetus. Also once the foal is born, if the mare is not a good mother, will not allow the foal to suckle freely or does not give it the moral support and education it needs, it will turn out perhaps weak and stunted or have social or psychological problems.

Conversely, a healthy mare usually breeds a healthy foal. Her attitude to life is also 'catching' as far as the foal is concerned; for example, if she is hard to catch the foal usually will be, too. If she is kind towards people, easy to handle and trusting, she will teach this to her foal. If she is high in the social hierarchy of the paddock, so will her foal be within its own peer group. The same goes for mares which dislike humans for whatever reason, are distrustful, kick, bite and so on. So from this point of view the dam has a bigger influence on the foal than the sire.

The reproductive organs

The mare's reproductive organs consist of the vulva or opening to the vagina and the clitoris, the vagina itself, the cervix (neck of the womb), the womb or uterus and the two ovaries with the Fallopian tubes running from them.

Roughly kidney-shaped, the ovaries vary in size at different times of year and may average about 7cm x 4cm (2¾in x 1½in). They are placed just under and slightly behind the kidneys. Each one contains many thousands of eggs, which are already present when a filly is born: she acquires no more eggs or ova during her lifetime but there are obviously far more than enough for her purposes, even if she has a foal every year during her breeding lifetime, which few mares do.

The eggs are surrounded by fluid-filled sacs or follicles which protect them and which swell considerably before the mare ovulates. This is when the ovary releases the egg from the 'ripest' follicle, known as the Graafian follicle, through the dimple on its surface which is called the ovulation fossa. It is these enlarged follicles which the vet feels (palpates) when he or she inserts an arm into the mare's rectum to examine her internal organs through its wall.

Running from each ovary is the Fallopian tube, a coiled structure about 20–30cm (8–12in) in length which carries the egg from the ovary. The stallion's sperm swim up these tubes to rendezvous with the egg which, if one of the sperm succeeds in fertilising it, travels down the tube into the uterus for further development.

The uterus itself consists of a main body with left and right horns, into each of which runs a Fallopian tube from an ovary. The uterus, Fallopian tubes and ovaries are suspended from under the spine by means of a membrane of connective tissue. The womb has muscular walls and a specialised lining, the endometrium, containing glands which secrete hormones to help to maintain and progress pregnancy.

At the opposite end of the uterus to the ovaries is the muscular neck called the cervix. This is tightly closed most of the time, only opening during parturition (foaling) or when the mare is in oestrus or season to allow the entry of sperm. The cervix opens into the vagina, a muscular tube which itself ends in lips, the vulva, which

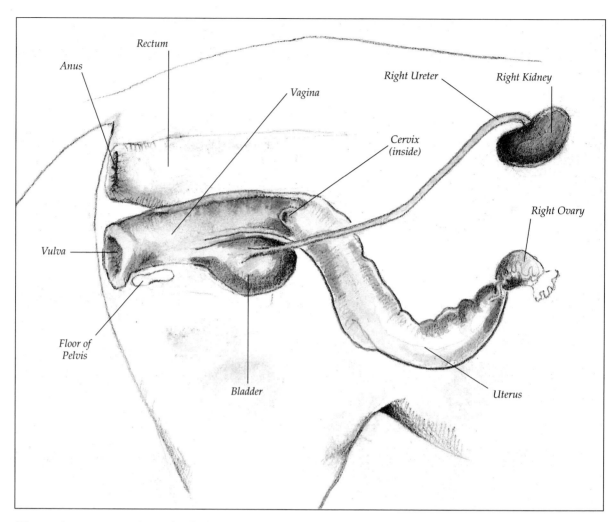

Diagramatic representation of mare's breeding organs.

act as a seal, helping to protect the vagina and uterus from the entry of disease organisms. In maiden mares, there is a membrane called the hymen which also helps prevent the spread of disease to the cervix and uterus.

The reproductive organs and the nearby bladder (which opens via a tube into the vagina for the voiding of urine) are surrounded by a bony ring called the pelvic girdle, which consists of the sacrum (part of the spine or vertebral column) at the top and the pelvis. The rectum also passes through this girdle, which is a convenient arrangement because it facilitates examination of the ovaries and uterus. The foal must pass through this ring during birth and it is the main reason why a wrongly positioned or presented foal can cause such serious problems at this time: the pelvic girdle cannot 'give' like soft tissue to allow the foal leeway to squeeze through if it is wrongly presented, although the ligaments in this area do soften at parturition to help facilitate the process.

Photoperiodicity and the oestrus cycle

The oestrus or 'heat' cycle of the mare is mainly controlled by day length on a year-round basis. It is partly integral to the mare, but warmth and feeding, particularly the amount of protein in the diet, also have an effect. Within the annual cycle of heats (or seasons) and the lack of them is the roughly three-week cycle which normally occurs from spring through to autumn – wheels within wheels, as it were.

Over millions of years of equine evolution, those equidae (horses and their relatives) which gave birth to their foals in spring and summer, when their natural food of grass and leaves was of its highest quality (young and high in sugar and protein) and at its most plentiful, had the highest chance of perpetuating the species. Foals born too early or too late suffered from adverse weather and limited food supplies. The colder weather and shorter days limited the growth of food, mainly grass, which itself reduced the milk the mare could make. Therefore, foals born at the wrong time did not have as good a chance of survival: they would be poorly nourished or might even starve and the cold would stress them greatly, perhaps even making it impossible for them to maintain their body temperatures (difficult for young foals, anyway), so that they died of hypothermia.

This annual cycle of heats occurring in spring and summer is still largely followed by our domestic horses, but it is not unusual for mares to continue to come into season, if only mildly, in winter. If they are well sheltered in winter, well fed, possibly rugged up and generally protected from natural winter conditions they do not sense full winter weather and so may appear to continue to cycle normally, although the seasons may not be fertile ones.

The four turning points of the natural year which determine the annual and, therefore, the three-weekly cycles of photoperiodic species (those intensely affected by day length) – of which the horse is obviously one – are the vernal (spring) and autumnal equinoxes, when daylight and darkness are of equal length, around 21 March and 21 September in the northern hemisphere; and the summer and winter solstices which are, respectively, the longest and shortest days of the year and fall around 21 June and 21 December. These times were vitally important to earlier societies, their animals, their crops and their relationship with the natural world; they were high days, celebrated by feasting, sacrifices (human and animal) and other gifts to their gods, orgies, fairs, processions and religious ceremonies connected with life, death, rebirth, passing to other lives and breeding and fertility. But even if the solstices and equinoxes now pass unnoticed by modern horse owners

Even early in the season, mares will start showing signs of being in oestrus – frequently passing small amounts of urine and mucus, 'winking' (opening and closing) the vulva and straddling as a sign of being ready for mating. Early 'seasons', however, are rarely fertile. (Vanessa Britton)

(they are often not even listed in diaries these days), they will notice the various changes they bring in their horses.

After the winter solstice, often by early February, mares usually begin to behave like 'brazen hussies' and start to come into season, although they will not yet be fertile as ovulation is not occurring. They squeal, kick and act silly; in fact all horses sense the turn in the year and begin to get restless. Winter coats start to cast often very soon after the solstice, and horses sense that 'the sap is rising'. If they have been mainly stabled during the winter, and apparently fairly content with minimal winter grazing, they are no longer! They seem to be able to smell the grass and cannot wait to get at it.

After the spring equinox, the day length is over twelve hours per twenty-four and it is this which triggers the onset of true oestrus cycles. The natural mating season gradually goes into full swing, the spring grass with its high sugar and protein content grows more and more, summer coats continue to come through and mares start cycling and ovulating normally roughly every three weeks.

Once the summer solstice has passed, things start to wind down very gradually. Summer coats begin to cast and by late August or September they have lost their bloom, hence the old saying, 'No horse looks well at blackberry time'. Mares continue cycling, having slightly less fertile seasons, until the autumnal equinox, after which there is less than twelve hours' daylight per day. The seasons then tend to become longer and less distinct, as in the spring, and the winter non-breeding period of the year begins. During this time, mares are said to be in anoestrus ('an' meaning 'without', so, without oestrus or not cycling).

Soon after the winter solstice, however, we notice once again that our horses have already begun to cast their winter coats and the year has come full circle. (The full coat change, of course, takes several months and horses keep their full coats in either summer or winter seasons for a very short time. They cast a little, then grow a little, then cast a little more and so on until the change is complete.)

Hormonal control

These external changes are indicators of what is happening inside the horse. They result from the gradual production of hormones in the body which put the horse into or out of breeding mode and apply to stallions as well as mares.

The complex interactions of the hormones involved in the reproductive cycle are not fully understood, but the following simplified, general picture may be interesting. For convenience, let us take the winter solstice as the start of the annual cycle. It is Christmas Eve on a stud and whilst the humans have been rushing around preparing for Christmas and arguing about who is going to be on duty on Christmas Day, the horses' brains have detected that the year has turned and the days are, barely perceptibly, becoming longer. Spring is on the way! Light rays enter the pupil of the eye and, in light-sensitive species, a message passes down the optic nerve at the base of the eye to the brain, which senses that the days are lengthening. It sends out chemical messages, hormones, to gradually gear up the body for the mating season.

The pineal body, a small gland in the mid-brain, secretes a hormone called melatonin, which depresses breeding activity. The darker it is, the more melatonin is produced and the less the horses feel like breeding. Light, especially bright light, is a melatonin suppressant, so the longer the days and the more light the horses experience (up to a limit), the less melatonin is produced and the more the horses gradually feel like breeding.

The hypothalamus, part of the lower brain, senses the decreasing amount of melatonin which stimulates it to produce gonadotrophin-releasing hormone (GnRH). GnRH in turn stimulates the pituitary gland, a vital gland just under the brain, to secrete follicle-stimulating hormone (FSH) which, as its name suggests, stimulates the development or 'ripening' of follicles in the ovaries. As the follicles grow, they themselves secrete the hormone oestrogen which causes the mare to show the familiar typical behaviour of a mare in season. The mare is now in oestrus, a state which lasts four or five days.

Oestrogen has a negative feedback effect on the pituitary gland, causing it in turn to wind down the production of FSH. Oestrogen also stimulates the production by the pituitary gland of another hormone called luteinising hormone (LH); this causes the rupture of the Graafian follicle described earlier and a reduction in oestrogen levels. The Graafian follicle releases its egg from the ovulation fossa into the Fallopian tube, where it may or may not be fertilised by one of the stallion's sperm. Ovulation occurs between twenty-four and thirty-six hours before

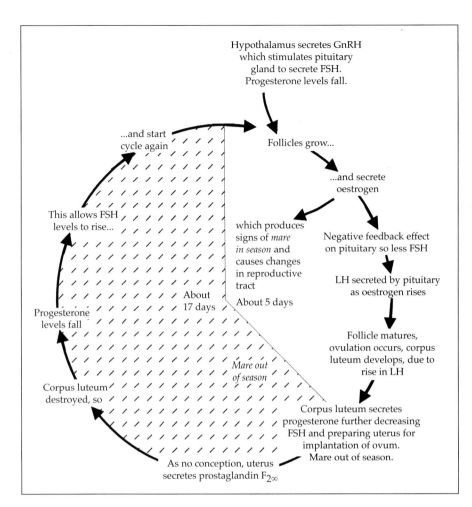

the end of oestrus.

The space in the ovary formerly filled by the follicle with its fluid and egg is now filled by a blood clot known as the corpus haemorrhagicum. Under the influence of LH, specialised cells (luteal cells) invade the blood clot and change it into a structure called the yellow body or corpus luteum. The corpus luteum grows and itself secretes another hormone, progesterone – and the mare goes out of season. She is now said to be in dioestrus, the period during the breeding season between heats. She is still cycling but is not in season at present from a practical point of view, not having a ripe follicle, and she will not accept a stallion. Dioestrus lasts about sixteen days.

With rising levels of progesterone and the decrease in oestrogen, there comes another negative feedback effect on the pituitary gland, causing it to reduce production of LH and once again secrete FSH. But before the mare can come into season again, the yellow body in the ovary has to be 'killed off' to stop it producing progesterone, because progesterone inhibits the in-oestrus/season stage of the cycle. The uterus now secretes yet another hormone, prostaglandin, for this purpose. This stops the yellow body producing progesterone; the hypothalamus recognises the lower levels of progesterone, promotes the secretion of GnRH again to stimulate the production of FSH once more, and another cycle begins.

This entire cycle takes about twenty-one days to complete, although there may be a very few days' difference in individual mares.

Conception

During ovulation, the egg in the Fallopian tube passes down towards its junction with the relevant horn of the uterus, where it must be fertilised by a spermatozoon from the stallion before it can pass into the uterine horn. Because sperm, normally, remain functional for longer than ova, the ideal time for mating to take place is usually about twenty-four to forty-eight hours before ovulation occurs.

If fertilisation occurs, the egg passes down into the uterine horn and, by the fifth day after fertilisation, finally arrives in the uterus itself. The mare is now technically pregnant, and the early maintenance of pregnancy depends on high levels of progesterone. (The corpus luteum or yellow body in the ovary has to keep producing progesterone until at least the seventy-fifth day of pregnancy. If it fails to do so, or if progesterone levels drop for any other reason during this period, the mare will abort.)

The mare's reproductive system has to be informed by the fifteenth day after fertilisation at the latest, by means of yet another hormonal message, that there is a fertilised egg, a conceptus, in the uterus. Although the exact train of events is not certain, it is believed that the conceptus itself produces oestrogens, the presence of which are sensed by the endometrium (the lining of the uterus) informing the mare that she has a fertilised egg in her womb. A further hormonal message is probably sent to the corpus luteum in the ovary, causing it to keep producing vital progesterone and so maintain early pregnancy.

If the vital message due by the fifteenth day is not received, the uterus will, as part of the normal twenty-one-day cycle, start to produce prostaglandin to stop the corpus luteum producing progesterone so that the mare can come into season again – and so it goes on.

Once the message is received, however, progesterone continues to be produced by the corpus luteum even if the egg fails or is lost for some reason after the fifteenth day. The mare will then remain in prolonged dioestrus and can be very difficult to bring into season again.

Artificial manipulation of the annual breeding cycle

The advantages of breeding horses according to their own and the earth's natural annual cycle have been explained. Unfortunately, Thoroughbreds and a few other breeds have their official birthdays designated by man as 1 January of the year of birth. This date was introduced in 1833 for reasons of economics by the Jockey Club of the United Kingdom (which controls the registration of Thoroughbreds). It was previously the much more logical 1 May.

Owners, who paid the bills, wanted a quicker return on the money spent on breeding, rearing and training their racehorses. As two-year-old racing became popular in the eighteenth and nineteenth centuries (again because owners were impatient to see their produce on the racecourse, trying to win them money and glory), it was realised that early foals matured sooner (within reason). They looked bigger and stronger when the time came to sell them as weanlings or as yearlings the following year, and so fetched higher prices. If a breeder kept his stock to race, again the earlier foals seemed to have an advantage in two-year-old races.

It was, therefore, decided to make the registration date 1 January instead of the much more sensible compromise suggestion of 1 March. The nearer to this date foals are born, the more time they have to develop strength and size. By the time the animals are mature (which, in horses, is not before five years of age and often later), any early advantage is lost, of course, but, in flat racing, most animals are retired from the sport well before then. The emphasis in this sport is almost entirely on two- and three-year-olds, with some animals remaining in training at four. There are few older flat racers. (Steeplechasing is different. In this sport horses are older and the age difference early in life is nothing like so important.)

Whatever breed or type of horse or pony you want to produce, if, for any reason, you need to arrange for foals to be born early in the year, it is necessary to shift forward the breeding season so that it begins at least two months before the natural season. As an example, the official Thoroughbred breeding season in the northern hemisphere begins in mid-February (mid-November in the southern hemisphere) and, as the gestation period of mares is around eleven months (340 days on average), foals start being born in January – although most are born later. Unfortunately, if they are premature and are born in November or December of the previous year, on 1 January they become officially a year old, so all advantage is lost.

Various practices have been devised to fool horses' brains into 'believing' that the natural breeding season is on the way so that mares will start cycling and stallions will also get 'in the mood'. Show horses of all breeds, types and ages destined for early shows are required to look as though they are in summer condition for the so-called pre-season shows, which start mainly in March, with some even earlier. It is not only breeding stock which respond to having the seasons of the year manipulated to move spring forward; it works with all horses, even geldings!

To promote breeding/summer condition whilst it is still midwinter, do not do as is usually done in the showing world and start overloading your unfortunate animals with rugs – although judicious rugging up plus a careful increase in feeding and protein levels helps. Instead, follow the lead of the more progressive Thoroughbred studs and, from the winter solstice, start exposing them to sixteen hours a day of full light, natural and artificial. Studs selling foals and youngstock in the autumn and wishing to retain summer condition can commence this light treatment gradually after the summer solstice so that the animals' brains do not detect a lessening in daylight.

To employ light treatment, you will need to invest in some full-spectrum light bulbs or strips. Natural daylight is full, white light comprising a spectrum from red to blue – red, orange, yellow, green, blue and violet (some people like to add indigo between blue and violet). If all these colours are painted in sections on a disc and this is spun on a nail through its centre, it will appear white. Full-spectrum lights are used commercially to encourage plant growth and also in human hospitals to treat 'winter blues' or Seasonally Affective Disorder (SAD). You should be able to order them through any good electrical shop. However, stress that you want full-spectrum lights, not merely those described as 'white light', which often simply means that the tube itself is white as opposed to a pastel, fashion colour. Ordinary light bulbs also work if they are at least 100 watts – ideally 200 watts – but they are not as good. Blue-spectrum, fluorescent lights, the most commonly available sort, do not seem to work at all; they also seem to cause headaches in humans and irritable behaviour in

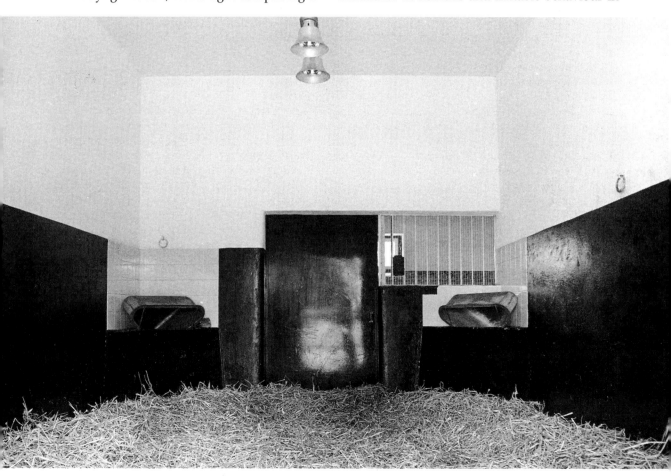

A five-star luxury box on a top Thoroughbred stud. Both mares and stallions can be encouraged to come into season earlier in the year than normal by keeping them warm, exposing them to extra daylight and feeding them well. This box has heater-lights in the ceiling and is big enough to give plenty of room without draughts. (Peter Sweet)

19

horses (perhaps also due to headaches) and are difficult to see by. The constant very rapid flicker given off by cheap fluorescent lights may not be discernible but seems to clash with the brain's natural electrical wavelengths and can cause depression as well as headaches, and a general feeling of being below par.

You can extend your horses' daylight hours by putting on the lights in the early morning and in the evening, or whatever suits your routine, so long as their exposure to light, including natural light, totals sixteen hours a day. For the technique to work, the horses must also experience eight hours of darkness.

It has also been found that just one hour's exposure to light between 2a.m. and 3a.m. has the same effect as the longer exposure detailed above. Obviously you would need a timer switch to use this method.

Your veterinary surgeon can also help advance your mare's breeding season by means of appropriate hormone injections from the beginning of February: this is quite a common practice.

The combination of light, hormones and, to a lesser extent, warmth and feeding will result in most mares being brought on to a regular cycle much earlier in the year than if it were all left to nature.

The behaviour of mares during the cycle

Mares often cause consternation among horse owners who are not planning to breed because of their 'mare-ish' behaviour. They complain because their minds are not on their work, their temperaments change (normally behaving in the opposite way to normal). They may harass geldings and other mares in the paddock and there may be mock harem battles for the attentions of any geldings or even high-ranking mares. They may even make approaches to their human attendants, failing other willing horses. All this, plus the disinclination to work, annoys some owners of working mares.

It is, of course, all part of the natural scheme of things. Mares in oestrus may become more affectionate and seek the close company of geldings or friendly mares. This may become flighty and playful near other horses, nipping and rubbing against them or standing with them, resting their heads over the other horse's neck or back. They will make soft 'cow's eyes' at geldings, nicker quietly to them and appear submissive even if they are normally quite self-confident and independent.

An in-season mare will also make frequent attempts to urinate and will pass small amounts of urine along with mucus. The vulval area and vagina will become reddened, moist and a little swollen. The mare will stand with her tail towards any apparently interesting gelding, raising her tail, straddling her hind legs and everting the lips of her vulva to expose her clitoris, a typical action called 'winking'. She is most receptive towards the end of her season and most efficient studs will try or tease mares, either with the stallion or with a stand-in called a teaser which will do the donkey-work of sounding her out but not actually get to mate with her. This takes some work away from the stallion but not everyone agrees with the practice, maintaining that it is very unfair on the teaser, which may not even be allowed a few mares of his own. Also, the mare, having been courted by and become very interested in one particular stallion may suddenly be confronted by a quite different one which she may not fancy nearly so much! Some breeders have maintained that this actually results in lower conception rates than if the mares are teased by their actual 'intended'.

Mares not in season are not receptive to a stallion. In a paddock, they will not make overly affectionate moves towards any other horse and will concentrate on the vital job of eating, the thing horses evolved to spend most time doing. A mare presented to a teaser or stallion when not in oestrus will put her ears back, with an irritated or even angry expression on her face and drawn-up nostrils, and maybe even bared teeth. She may attempt to bite him and will thrash her tail. If he nuzzles her, she will kick out quite violently.

Even when in season, however, some shy mares may exhibit none of the typical in-season behaviour. Although stud staff may believe such a mare is in season, the best judge of her condition is the teaser or stallion, which can tell, by the specific pheromones or scents she gives off, whether or not she is ready for mating. Very tactful handling and teasing of such mares is needed if they are to accept the stallion, even when fully in season. Such a mare may be a maiden (never having been mated before), have had a frightening or painful experience during a previous mating, may simply be shy or may have a low libido (sex drive). But careful timing, expert handling and an experienced, tactful stallion usually win the day in the end.

Chapter 3
Reproductive Function
in the Stallion

It is often not appreciated how important the health and condition of a stallion are to his breeding performance. Even today, with our improved knowledge of physiology and metabolism, one still sees stallions which are fat, not only in the showring but also at stud. Although a stallion may lose condition (bodyweight) towards the end of a busy stud season, this is not a valid reason for his being made overweight earlier on. A leaner stallion may well be healthier overall.

Obese stallions may be less fertile and energetic than those in good condition and, it follows, probably less healthy all round. For almost any horse, a good, healthy condition means that one should be able to feel its ribs quite easily but not see them. Performance stallions may appear leaner than this but they will probably actually be in a fit, muscular condition with the back couple of pairs of ribs visible. It does depend on the individual, however. If a horse starts the stud season in this sort of condition, has an ample diet, a calm environment, enough time out relaxing on good grazing and quiet, firm handlers he knows and trusts, there should be no reason why he should lose significant bodyweight as the season progresses.

Healthily taxing exercise is one commodity most stallions are very short of, however. In the UK, Ireland and parts of North America, many stallions, particularly those in the Thoroughbred industry, are only exercised by being walked in hand, perhaps lunged or turned out in their paddocks. This is not adequate exercise for a naturally athletic animal like a horse. To become and stay fit, the body has to be stressed, and adapt to that stress by strengthening and becoming fit. Leading out in hand – or lungeing which should not be overdone with any horse, anyway – will not promote this process effectively. There is a good case to be made for stallions to be ridden or driven several days a week by those competent to do so, and some studs now employ stallion riders specially for

this job. In addition to improved physical condition (and we are not talking about getting horses racing or competition fit, of course, just healthily athletic), this adds another aspect to a stallion's routine and promotes a balanced mental outlook and an improved temperament, which will benefit the other areas of his life.

The reproductive organs

The reproductive organs of a stallion comprise the testes, the epididymis, the vas deferens, the accessory glands and the penis.

The two testes or testicles (the size of which, research now indicates, does relate to fertility – the bigger the better!) hang between the thighs in a sac of skin called the scrotum. They are proportional to the size of the horse; in a Thoroughbred horse they are about 10cm (4in) long by 6cm (2½in) deep by 5cm (2in) wide. They are of a similar size to each other and are firm, smooth and regular in shape.

The scrotum is divided into two parts, one for each testis, lined with protective tissue called the tunica vaginalis. Between this and the scrotum is a lubricating fluid which enables each testicle to move comfortably inside its compartment. The testes are ovoid in shape and slightly flattened from side to side. Their lower part is free inside the scrotum whilst the upper part is attached to a membrane containing the epididymis.

The testes must be cooler than the body temperature for the sperm to be produced successfully, and are therefore housed in and protected by the scrotum and positioned fairly safely outside the body between the stallion's thighs. The cremaster muscles in the spermatic cord are temperature-sensitive. They relax when the stallion himself is relaxed or when the outside temperature is warm, allowing the testicles to hang down, and contract, pulling the testicles up closer to the body, when the stallion is stimulated or when the outside temperature is cold.

Internally, the testes contain a convoluted

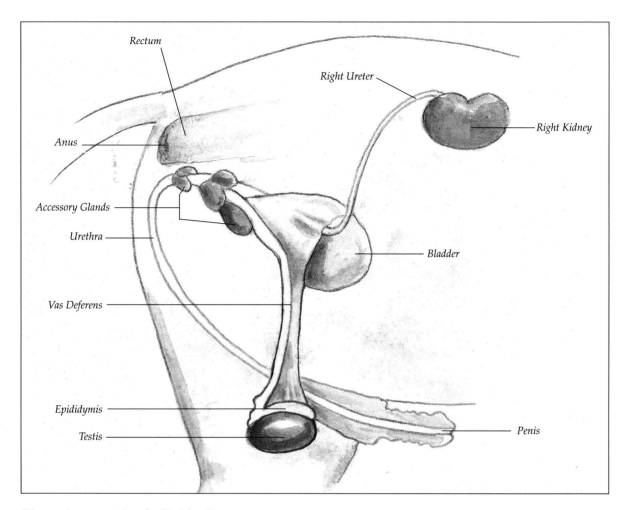

Diagramatic representation of stallion's breeding organs.

network of tubules called the seminiferous tubules in which the spermatozoa develop. The tubules also contain the cells (Leydig cells) which produce the male hormone testosterone. There are blood vessels and also connective tissue which makes each testis more stable by dividing it internally into small lobes containing the seminiferous tubules.

The testes are normally in the scrotum and outside the body at birth or soon afterwards. They are inside the abdomen *in utero* and then pass down the inguinal canal (a passageway between the abdomen and the scrotum) into the scrotum to hang outside the body.

When one or both testicles remain up in the abdomen after an age at which the colt should have reached puberty (around three years), the animal is termed a cryptorchid or 'rig' and may often be mistaken for a gelding. Rigs are, in practice, as sexually capable as stallions and behave accordingly, and they can cause significant management problems. A veterinary surgeon can have a blood test carried out to check the sexual status of a suspect animal.

The epididymis is a coiled tube (each testis having its own) with a head, a body and a tail and contains smaller tubules. It lies on the top border of the testis and its tail joins the vas deferens. The epididymis collects the immature sperm from the testis at its head; as they travel along it in the tubules they develop, and are more or less mature by the time they reach the tail of the epididymis, where they are stored before ejaculation.

The vas deferens, which continues on from the tail of the epididymis, carrying the sperm, runs, along with the spermatic artery, the spermatic vein and the cremaster muscle (together comprising the spermatic cord), upwards through an opening in the abdominal wall called the inguinal ring or canal. In the abdomen, it separates from the other spermatic vessels and turns backwards into the pelvic cavity, enlarging to form the ampulla, one of the accessory sex glands (the others being the seminal vesicles, the prostate and the bulbo-urethral glands), which produce most of the nourishing and protective seminal fluid in which the sperm travel. In this area, the vas deferens meets the urethra (the tube which carries urine through the penis as well as carrying the stallion's ejaculate).

The penis is normally flaccid and lies up in a protective fold of skin called the prepuce or sheath. The only times it is usually seen are when the stallion is staling (passing urine), when he should let his penis down, and when he is sexually excited. It is composed of erectile tissues which engorge with blood to create an erection during sexual activity.

The tip of the penis ends in the glans penis, which looks like a cap to the organ with a rim, through which the urethra projects past the end of the penis a few millimetres. The depression around the end of the urethra is called the urethral fossa and a greasy, greyish-coloured, unpleasant-smelling discharge called smegma often accumulates here, as in geldings, and inside the sheath. Dry smegma cracks and drops off naturally but most stallions and geldings benefit from having it removed with warm water and a mild soap to break down the thick grease, rinsing thoroughly afterwards. Horses differ as to how often this should be done but once a fortnight seems about right and it does seem to make them more comfortable.

Semen

Semen is the fluid seminal plasma plus the sperm. The sperm travel in this fluid, which is made by the accessory sex glands (the prostate and bulbo-urethral glands and the seminal vesicles) and the fluid is deposited inside the mare's genital tract from the stallion's penis during copulation.

Semen is white to pale grey in colour and slightly viscous in nature, and the nutrients and chemicals it contains enable the sperm to develop and live on inside the mare until fertilisation takes place.

Sperm can live in the mare for up to about thirty-six hours, but the body appears to have a mechanism whereby ageing sperm are killed off, probably by substances in the seminal plasma. This could be to prevent an old sperm fertilising an ovum, which seems to result in a higher incidence of defects occurring in the resultant foetus.

Semen quality is directly related to stallion fertility and most professionally run studs have laboratory checks carried out on their stallions' semen to determine its quality. The season of the year also has a considerable effect on this factor and stallions are just as much affected by light as mares (see chapter 2). Although a stallion is quite capable of serving mares all year round, most are much less interested in winter and their semen contains far fewer sperm than in spring and summer, which comprise the natural mating season. Light therapy will stimulate sex drive and the production of sperm earlier in the year and stallions who serve at stud all year round, being flown between the northern and southern hemispheres, appear to be able to continue serving mares with no ill effects or decrease in fertility provided they are getting their required daily amount of light.

Sperm

Healthy, normal spermatozoa (singular: spermatozoon) are very active 'self-starters', driven to swim strongly in the seminal and other fluids present in the mare's genital tract towards the Fallopian tubes, where their aim is to fertilise an ovum or egg. They are unimaginably tiny. An egg is about the size of a grain of sand and one sperm is about one-hundredth the size of that egg. In one normal ejaculate, there may be several thousand millions of sperm, only one of which is needed for the fertilisation of an egg! The volume of an ejaculate is 60–70ml (2–2½fl oz), containing 100 to 800 million sperm. Despite this, by no means every act of copulation results in a pregnancy.

A sperm consists of an elliptically shaped head, a middle and a tail with an endpiece. The head is shielded by a protective cap called the acrosome and contains the stallion's genetic material, which will combine with that of the mare in her egg to make the new individual. Deformities of the sperm, which are not uncommon, result in their not being able to swim strongly or at all, failing to go in the right direction or failing to pierce the egg when they find it. Fortunately, there are usually more than enough normal, healthy sperm to ensure the likelihood of success (50 per cent or more).

Hormonal control

Like mares, stallions are subject to the influence of hormones. Increasing day (or appropriate artificial) light stimulates the hypothalamus to secrete GnRH, which travels in the blood to the pituitary gland, causing it to produce gonadotrophic hormones – the same as those produced in the mare. In the stallion, however, LH is sometimes called interstitial-cell-stimulating hormone (ICSH). The gonadotrophic hormones urge the testicles into action, FSH stimulates the manufacture of sperm and LH or ICSH stimulates the production of the male hormone testosterone.

This plays a part in sperm production and in the development of the reproductive organs in the foetus, the descent of the testicles and the process of puberty. Very visible to onlookers are its other effects of inducing the stallion's sex drive and his interest in mares, his herd leadership qualities, the tendency of some to assert their authority over humans as well as other horses (often by nipping or actual aggression) his proud bearing and the well-developed crest to his neck.

The sexual behaviour of the stallion

In establishments where stallions run with their mares all year round – not a particularly common arrangement in most Western countries – and in wild and feral situations, they may well lose interest in sex outside the normal breeding season but they take a concerned, affectionate interest in their harems and youngstock all year round. As the breeding season approaches, feral stallions start to pick and choose which youngsters are to be allowed to stay in the herd. The pubertal males are the first to be kicked out but sometimes fillies are also excluded.

Stallions' ambitions do not run to acquiring as many mares as they possibly can. 'Wild' herds often consist of a stallion and only a handful of mares, although larger ones do exist. A stallion knows he can cope with only so much mating and so many rivals sneaking in and stealing or impregnating his mares. In some regions, large herds consist of a number of smaller or family units with a stallion and a few mares with their followers. When peace reigns, these units have been observed to mix freely but in times of danger or excitement they split off into their family components within the larger herd.

Although environmental pressures relating to grazing and water do not affect most domestic breeding herds, in feral conditions horses and other equidae have a finely tuned sense of how many animals an area will support. On a visit to a private collection of zebras and Przewalski horses on an estate in the British countryside some years ago, I was fascinated to learn how one of the Przewalski stallions would savage to the point of death any new mare introduced to his herd which took the number above six (they had 8 hectares (20 acres) of good-quality, lowland grazing on which to roam). If one of his mares were removed for any reason, he would then accept another. When this stallion killed a mare, the staff thought he had taken exception to her as an individual, but watching him carefully round the clock, they saw him attacking her replacement so they quickly moved in and took her away. When his daughters reached puberty, he also attacked them, although this does not seem to be uncommon in feral equidae.

The Przewalski stallions in this collection could not be mixed at any time of year, but the zebra stallions lived happily together outside the breeding season. A horse charity in the UK received several mature stallions which had been reared together and which got on perfectly well all year round.

An interesting phenomenon particularly noticeable in feral horses, and which I observed on one domestic stud which was not particularly carefully run, is that a new herd stallion will often kill off the foals of the previous year's stallion. It is believed that they do this partly to remove a predecessor's genes from the herd and partly to bring the mares back into season so that he can impregnate them with his own genes. As far as I can determine, the murderous tendency does not apply to a previous stallion's yearling colts which are already in the herd. Like his own offspring, however, they will normally be driven away at puberty. In domestic studs where stallions run with mares, it is worth bearing this trait in mind; it has been reported to me from several domestic sources.

Domestic stallions display just the same sexual instincts as wild and feral equidae, but in commercial studs they often live very artificial, controlled lives and are not allowed to behave naturally – although they do express a desire to do so. Animals intended to be kept entirely for breeding are often regarded as special and are well disciplined and handled from an early age to facilitate control later on. However, they cannot deny their natural urges, and most expert

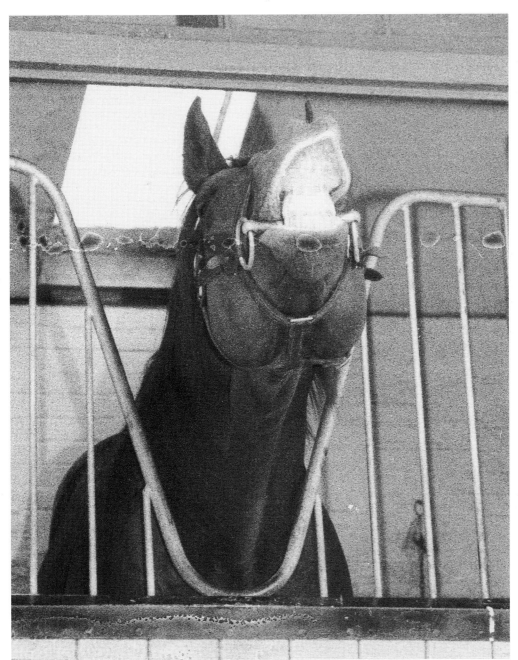

This stallion is performing the characteristic flehmen action which enables him to better savour the scent of a mare in season.
(Vanessa Britton)

stud staff and stud vets admit that they know better than any human when a mare is ready for mating, probably because of the pheromones or scents she gives off which change according to the stage of her cycle. A stallion is believed to be able to smell a mare in season, when the wind is in the right direction, over a mile away. He uses smell, taste and sight in his final judgement and all these stimulate him by means of hormones and his nervous system.

Because of the danger of kicks and broken bones to valuable and often much-loved animals, trying (whether by the intended stallion or a teaser) is normally done with the horses one on each side of a very strong, solid and usually well padded barrier called the trying board. The stallion will approach the mare (usually restrained by his handler) like any horse and the initial contact will usually be nostril to nostril. The usual squealing and nickering will occur and

25

the stallion, and probably the mare if she is not in season, will strike out with one or both forelegs, or at least stamp. The mare is then presented sideways on to the stallion so that he may nuzzle and nibble her along the neck, back and loins, and finally reach the root of her tail, when he will smell and lick her vulva.

He will then probably perform the characteristic flehmen action, in which the aroma of the mare is examined fully. Smells consist of tiny physical particles of matter, which dissolve in the fluids in the nasal passages. The stallion lifts his muzzle high and curls up his top lip, mainly closing off the nostrils, so that the fluids run towards a special sensory organ called Jacobsen's organ in the nasal passages. This is responsible for assessing the smell and so informing the stallion whether or not the mare is ready to mate.

By this time, the stallion's penis will probably be erect, even if the mare is not fully ready to mate, and he will snort and roar. However, he may not become quite so excited when a mare is not fully ready as he would with a mare full in season.

If the mare is ready, she will invite him by leaning towards him, raising her tail, opening and closing her vulva (called winking), straddling her hindlegs and dribbling urine and mucus. If she is not ready (and just because she is in season it does not mean she is fully ready to mate), she will probably squeal, put her ears back and thrash her tail, and may try to bite the stallion; she will almost certainly kick out high and hard to leave him, and the attendants, in no doubt that she is giving him the brush-off. Everyone will have to try again the next day.

Trying can often have the effect of bringing 'shy' mares – those who do not show much or at all when they are in season – into season. Some years ago, Ireland's National Stud ran some now-famous trials of the effects of sound on horses. In the mare barns, it was found that relaxing music, as opposed to heavy rock, made for contented mares. Playing the sound of a stallion calling to a mare, or courting or mating one, often brought into oestrus mares which had previously shown no interest and were not in season – and this was without them even seeing or smelling the stallion. This could mean that it would be advantageous to have covering take place at least within earshot of mares intended for mating, and maybe within sight, too, as mares running loose with stallions are known to take an interest in the male's attentions to other mares.

When stallions run with mares (that is, live out with them on pasture in the domestic version of a natural herd), stallions treat mares who are not in season as ordinary herd members and are usually friendly and affectionate. The mares behave similarly. When there is an in-season mare on the scene, however, the stallion often only has eyes for her and may spend all his time with her, copulating repeatedly until she goes out of season after a few days, by which time one of the other mares may be ready for his attentions.

Working and breeding stallions

Stallions which are properly handled and managed can quite safely hack out, for example, with other horses, even other stallions. In flat-racing yards, the various strings work together in their normal way, colts (entires) at the front, then geldings, then mares and fillies not in season with in-season females bringing up the rear. In this way, the entires (who are discouraged from any interest in sex, anyway) are not tempted by having in-season females in front of them wafting their pheromones under their noses, and not only making it impossible for them to concentrate on their work but possibly making life positively dangerous for all concerned.

Stallions with a dual role – breeding and working – quickly learn which job is on the agenda, not only from the people handling them and the clothes they are wearing but also from the tack that is used. Stallions being used for trying or mating should wear a strong stallion bridle with the same type of bit, or even the same actual bit, each time. The horse soon learns that this bridle and bit mean sex – or at least a bit of fantasising about it. If the horse is being shown, a different bridle, perhaps with a traditional straight-bar or ported stallion bit, can be used. Provided the horse is never used for mating in this bridle and bit, he will not expect sex when wearing it. Similarly when working – provided his work tack is used only for that and never for his stud role, he will not be primarily interested in sex. However, if an attractive mare, in season or otherwise, puts in an appearance he can surely be expected to show at least some interest, even though he will know for sure that courtship and sex are not on the menu. If he does not, particularly during the breeding season, you may be forgiven for thinking that there is something wrong with him!

Chapter 4
Genetics and Heredity

Every horse is a combination of its parents' genes. In the body cells (with the exception of red blood cells) are nuclei or 'control centres' which contain structures called chromosomes. Chromosomes are long threads made mainly of a famous substance called deoxyribonucleic acid (DNA), which carries the genetic code, a sort of 'instruction leaflet', for determining hereditary characteristics.

Genes are segments of DNA, sited in particular locations on the chromosome, and there are thousands of genes in each cell, each one carrying a specific hereditary trait. Genes control more or less everything about an animal: its appearance, its behaviour, its physiology, its biochemistry, its metabolism, its tendency in certain circumstances to develop behavioural characteristics not formerly apparent or to develop certain diseases and, in general, how efficiently it functions.

Different species possess different numbers of chromosomes in their cells. The domestic horse, *Equus caballus*, has sixty-four, but other equidae have different numbers. The Przewalski horse,

The production of twins often runs in families. Twins rarely survive to term and, when they do, one is invariably bigger and does better than the other.

Family likenesses are obvious in prepotent lines. These two well-developed foals are cousins and not only look alike (being by the same stallion out of full sisters) but have an affinity for each other above all the other foals in their group.

for instance, has sixty-six. Chromosomes are usually paired in the cell, so we speak of the horse as having thirty-two pairs, each relating to a particular characteristic such as colour, temperament, abilities, height and so on.

Some genes always pass on their characteristics and are called dominant; others, called recessive, will be present but their characteristics will not be expressed in the resulting foal if they are 'overridden' by a dominant gene. A recessive gene is only expressed if it is transmitted by both parents. If a mare, for instance, has a dominant gene for a particular characteristic and the stallion with whom she is mated has a recessive gene for it, the offspring will inherit the dam's characteristic,

and vice versa. Many people think of stallions as stamping their stock or always producing offspring like them (called being prepotent); they overlook the fact that mares have just as much chance of stamping their stock. The reason why more attention is usually paid to the prepotency of the stallion rather than the mare is that stallions can produce many more foals in a lifetime than mares and so have much more genetic influence on the equine population as a whole.

Passing on genes

There are some body cells which do not possess thirty-two pairs of chromosomes. These are the sex cells or gametes – the stallion's sperm and the mare's egg. These each possess half the normal number (thirty-two single chromosomes) so that when they combine during fertilisation the resulting zygote – the very beginning of a new life – possesses the full and necessary complement of thirty-two pairs, half from each parent.

The process of cell division or mitosis, by means of which the cell divides into two halves, each containing the normal number of chromosomes, is magical to watch under a microscope – the actual creation of a new individual. The cells divide and divide, becoming specialised and replicating their genes again and again, until the embryo becomes a foetus and the foetus becomes a full-term foal ready to enter the world as a genetically unique individual in its own right – and, unless we interfere, ready ultimately to pass on its own genes to the next generation.

Genes have obviously been passed down for millions of years. However, it is generally felt that the more distant an ancestor the less influence he or she will have on a foal. Some genes will of course come down from a particular ancestor but there will be many others which come from other ancestors, so individuals are always just that – individual, a mix of ingredients which ensures that one never gets exactly the same animal twice.

Some years ago, I saw a lecture-demonstration involving a descendant of one of the greatest racehorses and sires of all time, St Simon (which was foaled in 1881). The descendant had never achieved his ancestor's success on the racecourse (although I believe he did win some minor races) but I and another spectator, being followers of racing, were very familiar with photographs of St Simon and both gasped with astonishment when this horse was brought in. He was the absolute image of his ancestor, so much so that looking at him felt like going back a hundred years! In fact, when the horse appeared, we did not know he was a descendant of St Simon and nor did his handler but, on checking his breeding with her, we realised that he was. Several very fast and successful racehorses belong to St Simon's line, but none of them looks like him, and I have never seen a descendant of his, famous or obscure, who resembled him as much as this horse did.

The point of this little story is simply to illustrate that the characteristics possessed by one individual are never passed down in their collective entirety to his or her offspring but are scattered among them. It is well known that a mare can be sent to the same stallion every year but the foals will never be identical and sometimes they can be quite different, according to which genes and characteristics happen to be inherited during that particular mating. If it were not so, wouldn't breeding be boring?

From the point of view of selecting the genes on which you want to base your foundation stock if you are planning to start a stud or line of horses, or even the one and only foal you plan to breed, your best plan is to look at as many of your mare's and stallion's offspring as you possibly can. Ask as many of their owners as you can reasonably trace about their physical and psychological characteristics and try to determine which traits seem to be coming from which parent.

Inbreeding and line-breeding

Desirable qualities or genes cannot only be selected and bred for but also fixed in a family by mating together genetically related individuals (those related by blood) which show or are known to regularly pass on those characteristics.

Inbreeding is normally regarded as the mating together of very close relatives such as dam and son, brother and sister or sire and daughter. This rather extreme practice has produced some phenomenally prepotent stock, not only in horses, but it can be very risky, as undesirable characteristics can just as easily be fixed in the family as desirable ones. It creates a build-up of genetic dominance (called homozygosity) in the herd, meaning that those animals will breed true for (always produce) those chosen characteristics. However, as well as fixing undesirable traits, this can also eventually lead to a loss of vigour and performance in the stock.

One advantage of inbreeding is that it can bring out the presence of poor qualities passed on by recessive genes and thus not normally expressed or visible. This enables a skilful breeder to eliminate them from his or her stock.

Mating half-siblings, aunts, uncles and nephews and nieces reduces the effect of inbreeding and mating full cousins reduces it further.

Line-breeding is the mating together of individuals which have a common ancestor somewhere back in their pedigrees, usually just a few generations back. This, too, is an effective way of intensifying a desired characteristic and is not so risky as inbreeding. Done throughout a herd or family, repeatedly crossing different individuals with those which have a particular animal or its close relatives a few generations back in their pedigrees, line-breeding can also eventually result in a build-up of homozygosity.

At this point an 'outcross' is said to be needed – the mating together of individuals completely unrelated. Called outcrossing or outbreeding, this produces the well-known phenomenon of 'hybrid vigour', a hybrid, technically speaking, being the offspring of two genetically different lines. This is impossible in most registered breeds, of course, since formal, registered breeds are usually all based at the outset on a few individuals (maybe of different breeds or types themselves). So although a breeder may cross two animals with no common blood ancestor for several generations back, it is still line-breeding that is being carried out.

The Thoroughbred is probably the best example in the world of a breed tracing back to only a few individuals. It is well known that all Thoroughbreds trace back to at least one of three oriental foundation sires – the Darley Arabian, the Byerley Turk and the Godolphin Barb (or Arabian). As another example, all Morgans trace back to just one stallion, Justin Morgan. A true outcross is becoming more and more difficult these days as breeds continue to be created by mixing existing breeds.

The rapid advances in DNA 'fingerprinting' or identification for horses, which builds on the blood-typing of individuals and breeds done in previous decades, plus the prospect of an Equine Genome Project which aims to map the genes of every breed and type of horse or pony in the world, will open up a huge Pandora's box of knowledge and questions about the origins of many a breed or individual. Indeed, this is already happening with the practical application of DNA identification. It is an advance which will surely change our understanding and perception of where our horses come from and what we really have grazing in the paddock outside.

Chapter 5

Pregnancy

Conception, and therefore pregnancy, begins when a male sperm, in company with many others, having swum up the oviduct or Fallopian tube and located a 'ripe' egg, pierces and enters it. The genetic material then mixes, as described in chapter 4, to provide the full number of chromosomes needed to create a new zygote, or conceptus, as this combined organism is called. Sperm seem to need the competition of other sperm to encourage them to reach a viable target and make contact, but once fertilisation has occurred the egg becomes resistant to piercing by other sperm. Two spermatozoon cannot fertilise one ovum.

The conceptus, now continually undergoing cell division and multiplication, takes six days to travel down the Fallopian tube and reach the uterine horn running from the end of it. The T-shaped uterus has two horns or arms, the vertical stroke of the T being the body of the uterus which is sealed during pregnancy and when the mare is not in season by the tightly closed cervix. The conceptus, as it is now called, is mobile and travels around the uterus at first, not starting to attach itself inside a horn (not necessarily the one on the same side as the ovary which produced the egg) until about seventeen days after service and becoming firmly attached by the twenty-fifth day.

The cells grow and specialise according to the role they will play in the mature individual – brain cells, blood cells, nerve cells, liver cells, muscle cells and so on. By the thirty-sixth day the foetus, as it is now called, has actually adopted a recognisably equine shape and is showing the beginnings of internal organs and legs.

Pregnancy diagnosis

Most breeders understandably want to know as soon as possible whether or not their own or a visiting mare has conceived and there are various methods of diagnosing pregnancy. One old, practical way is to present the mare to the teaser or stallion at the time when she would normally be in season again if she had not 'taken', and observe the reactions of the two of them. If the mare acts as though she is not in season and the stallion is not particularly interested anyway, she could well be pregnant. This basic method is still the only one used by many amateur breeders.

A more scientific method is rectal palpation, in which the vet inserts a lubricated and plastic-gloved hand and arm into the mare's anus and palpates (feels) her ovaries and uterus through the wall of the rectum. This method can be used about three weeks after service, and the vet can either give a definite diagnosis or a probability opinion, according to what he or she detects.

You will have to wait longer for a blood test, which detects the presence or absence of the hormone pregnant mare serum gonadotrophin (PMSG), more usually called equine Chorionic Gonadotrophin (eCG) because it is now known to be produced by the chorion, the outermost membrane of the embryo itself which eventually becomes part of the placenta. The blood test can be done between forty-five and ninety days, and it has the peculiarity of positive results being almost certainly accurate from forty to 120 days but negatives possibly being inaccurate.

From 120 days to full term, a urine test can be done. In this case oestrogen is the hormone looked for. Urine tests are usually very accurate.

But probably the most reliable test in general use today is the ultrasound scan. It is used for several purposes in veterinary and human medicine but, from the point of view of equine pregnancy diagnosis, it can give the earliest result yet, being able to detect a developing embryo from two to three weeks after service. This test is extremely accurate, but it is still not infallible.

With an ultrasound scan, a lubricated probe which emits sound waves (above the frequency discernible by the human ear) is inserted into the mare's anus. It is positioned and moved carefully around over the uterus. The soundwaves pass through the rectal and uterine walls, contact the embryo and return as echoes which are displayed on a monitor or screen.

During both manual palpation and ultrasound scanning, some mares become upset and some vets and breeders feel that their distress can contribute to the abortion of their embryos, later if not immediately. However, many mares accept the procedures without fuss.

Another advantage of ultrasound scanning as a means of detecting pregnancy is that it can often detect twin embryos at this early stage, giving the vet a chance to abort one or both of them; horses are not well equipped to carry twins successfully to full term. On an ultrasound scan, however, an endometrial cyst on the lining of the uterus may easily be mistaken for an embryo so the vet may advise you to have your mare scanned before service to detect the presence of cysts which may confuse a later scan. Sometimes cysts can also interfere with conception. Early pregnancies can also be confused with ovarian follicles.

Ultrasound scanning is a fairly expensive operation and several scans, up to about thirty-five days after service, and one before, may be

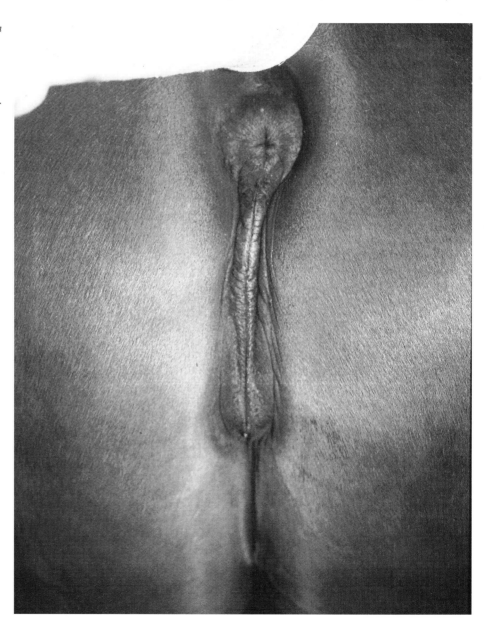

Good breeding conformation of the vulva. The vulva is straight rather than sloping back and the anus is not sunken, therefore droppings will not fall on the vulva and contaminate the vagina.
(Vanessa Britton)

recommended. However, you may only be able to afford one or two, in which case, the vet may advise you to have a pre-service scan done, followed by another four weeks after service, which will be very reliable.

Infertility in mare and stallion

If, despite all your efforts and those of the stud staff and vet, your mare is not holding to service (becoming pregnant), you have to consider why not. The staff and vet concerned will have a good deal of experience of this, as horses bred in domesticity are not particularly fertile compared with other species, it seems. Here are some of the reasons why things may not be going according to plan.

- You could simply be trying too early in the year without using the benefits of light or hormonal therapy. Although mares' early seasons may seem quite pronounced, they are not usually fertile because they will probably not be cycling regularly and will not be actually ovulating. Any hormonal abnormality which affects the normal reproductive cycle can prevent a mare (maiden, barren or recently foaled) conceiving, but can often be treated successfully.
- Mating at the wrong time in the mare's oestrus cycle does not give the best chance of conception. Careful note should be made of her seasons and of the stallion's or teaser's reaction to her when trying, in order to check that the stallion's services are not wasted by poor timing and that the mare is actually fully in season, with ovulation imminent. The ovum is only capable of being penetrated for about four hours after ovulation: as sperm take about five hours to travel up to the Fallopian tube, the mare should ideally be mated about twelve hours before ovulation or not longer than a day before (as ageing sperm are less 'able'), and obviously not afterwards. Ovulation itself commonly occurs about twenty-four hours before the end of oestrus, so it is obvious why very careful monitoring of the mare's normal oestrus cycle is vital to accurate timing of mating and, therefore, conception.
- Very young mares do not conceive as readily as slightly older ones. This is particularly true if they have been in hard training such as in racing, when it can take a year or so for them to let down, relax and become more physiologically 'normal'. When this happens they will probably conceive readily enough.
- Some mares which have proved extremely difficult to get in foal when mated in hand

conceive without a problem if allowed to run with the stallion. Such mares obviously need the more natural approach of courting, enjoyment, freedom and time to develop an attraction to and relationship with their suitors. Mares which are over-restricted (twitched, hobbled, having a foreleg strapped up etc.) during service may also be distressed or at least put off the whole process, quite understandably; and these practices have been described by several highly regarded and successful breeders as being equivalent to rape. Situations like these are reported by a great many people, and figures reported seem to confirm it. It is an aspect of breeding which should receive more serious consideration.

- The conformation of the vulva may be increasing the chances of reproductive tract infection. The vulva should be vertical and not protrude beyond the anus; if it does, every time the mare does a dropping, faeces will contact the vulva and infection can certainly spread inside the vagina and up to the cervix and uterus.
- A mare with a 'loose' vulva, where the lips do not meet closely (either due to basic faulty breeding conformation or to having had a lot of foals), tend to suck air into the vagina (they are often called 'windsuckers', not to be confused with the stereotypy of the same name, where the horse sucks air into its throat). Germs can be sucked into the vagina and, again, cause infection.
- In mares with poor conformation of the vulva and vagina, and possibly older mares, where the vagina sags down a little, urine can pool here and encourage the development of infection or adversely affect the biochemistry of the tract, neither of which promote the health and vigour of sperm. Most vets would advise not breeding from such mares.
- The mare may be simply not a very fertile individual. Over the generations, we have used more and more artificial methods of getting mares in foal when they may not have conceived otherwise. This means that sub-fertile mares among your mare's ancestors may have been induced to produce foals and passed on their tendency towards infertility to their offspring until it causes real problems in your own animal. Such mares should also not be bred from.
- The mare's general health and nutrition may mean that she is simply not in good enough physical condition to conceive – and being too fat is as disadvantageous as being too thin.
- The mare may have a problem with her uterus. She could have an infection which your veterinary surgeon can check on and probably

treat. Recent studies have shown that a mare with fluid such as pus or mucus in her uterus is only half as likely to conceive as one with a normal uterus, and is more likely to abort.

- A crucial factor in the efficiency of any broodmare is the condition of the endometrium (the lining of the womb). Professor W.R. ('Twink') Allen of the Thoroughbred Breeders' Association Equine Fertility Unit, has carried out pioneering work on, among other things, artificial insemination and embryo transfer (see chapter 10). He told me: 'Once your mare is diagnosed as having age-related degenerative changes in the endometrium, you must be firm and retire her from stud, no matter how brilliant a record she may have. It is useless giving her a year off, as is commonly done – it doesn't do any good.' The state of your mare's endometrium is something your vet can check by means of a biopsy (inserting a special instrument into her uterus and clipping off a tiny piece of its lining for microscopic examination).

- Ovarian tumours (granulosa cell tumours) can present problems as they cause the affected ovary to secrete oestrogen, inhibiting the pituitary gland and preventing normal cycling. Mares with this problem often appear very hyped-up, difficult to handle or even aggressive and although hormonal therapy can be tried the usual route taken is to remove the affected ovary, normally only one being involved.

- Old mares may naturally find it harder to conceive than younger ones, regardless of the state of the endometrium. In this respect, a study indicates that broodmares are at their best, being most fertile and producing their best foals, from the age of about nine to twelve or thirteen years of age. It was suggested that, apart from their chances of conception being highest between those years, it would pay breeders to send such mares to the best stallions possible during that period. This would maximise their chances of getting the world-beaters most breeders want and getting the best of both the mare's and the stallion's input. Older mares can still be used, of course, but perhaps there is a good case, especially in view of increasing endometrial degeneration as mares age, for retiring broodmares from stud earlier. Rather than persisting in trying to get them in foal year after year, they can be returned to work as, say, family hacks or hunters. I have known several older mares who positively thrived on returning to work and showed every sign of not wanting to bother any more with stallions, or even youngsters in the same field. Enough is enough, it seems!

- If a mare which is new to you fails to conceive at her foal heat (the first season after foaling), try to find out whether she normally does this but conceives at the next heat. Many mares do not conceive at the foal heat and some private studs now do not try mares until the following heat anyway, believing that the extra few weeks gives the mare and foal time to settle and the reproductive organs a little time to recuperate. Some mares naturally do not have foals every year, and yours could well be one of these.

- Infertility or poor fertility rates in the stallion may be the reason for your mare failing to conceive and may be due to several reasons, including: poor handling; bad management; sperm abnormalities; chromosomal abnormalities (which can also affect mares); psychological factors such as pain during previous matings, perhaps from being kicked or by being dragged off a mare too soon; immaturity or a psychological barrier to sex (most solely working yards actively and strongly discourage any inclination towards sex which causes the horse to think that it is wrong and may result in punishment); over-use, which means the testes cannot keep up with the required rate of sperm production and the sperm do not have enough time to mature; infection; painful feet or hind legs due to poor farriery, injury or disease; weakness or pain due to old age or arthritis maybe from degeneration or an old injury; being expected to cover too many mares when very young or old; and a failure of the horse to ejaculate. These problems, while of interest, may not directly concern the novice owner of a broodmare and should be able to be put right by good stud staff and a good stud vet, or a decision made to retire the horse if necessary.

Checking progress

Even if your mare does conceive, she could abort at any time. This seems to be most likely early on in the pregnancy. Following ultrasound scanning, blood tests can be done to check that she is still pregnant and, later on, also urine tests. In addition, you should try to keep a check on the mare's environment to try to detect an aborted foetus (these have been known to lie around in long grass for weeks). Later on, the mare's gradually increasing belly is a good sign. The big belly carried by a pregnant mare is a characteristically different shape from one which is just the result of obesity; the bulk in pregnancy is lower and further back than if the mare is simply fat.

Pregnant mares also tend to change temperament, which can be a good sign that all is progressing as you wish. Mares which are usually stroppy, independent and uncooperative, for example, may become affectionate, interested in humans and kinder-natured – and vice versa. They may also become gradually less inclined to keep active.

One of the advantages of keeping a pregnant mare gently exercised is to prevent the level of obesity which may confuse the unwary or novice breeder into thinking she is pregnant when she is, in fact, fat from being turned out so that she can 'eat for two' and take life easy. Mares actually benefit from judicious work up to three or four months before foaling. If they are fit, toned up, mentally occupied and interested in life, they must have a better chance of producing a healthy foal, unless there is some reason why a vet has advised that they should not work.

Life in the womb

The average length of gestation or pregnancy in the mare is 340 days, with about twenty days' leeway either way. As prey animals, horses have to produce young which will be sufficiently developed at birth to be able to recognise their own dams, to be on their feet within about half an hour to an hour, to learn very soon after that where food (milk) can be obtained and to have the instinct to follow their dams wherever they go and at whatever speed. This means that a foal has to have a well-developed brain and senses, and its eyes have to be open, unlike predator mammals. Its legs also have to be almost as long as they will be in adulthood so that it has a chance of keeping up with its dam and the herd fairly soon after birth, should danger threaten. Young foals are far more independent than, say, puppies and kittens, which are dependent on their mothers for several weeks – and certainly more so than humans, which need almost total care for years.

The embryo/foetus develops by means of the cells of which it is composed continually dividing and differentiating, specialising into cells which will ultimately have specific functions. Small blood vessels appear very early, by the sixteenth day and by the twenty-third day the embryo is taking on an actual shape and producing the beginnings of limbs and internal organs. By the thirty-sixth day, it is recognisable, to the trained eye, as a horse. By the time of foaling, the foetus has its coat and some mane and tail hair and its body is fully formed and functioning, to the extent that it can survive in the outside world with its dam's protection.

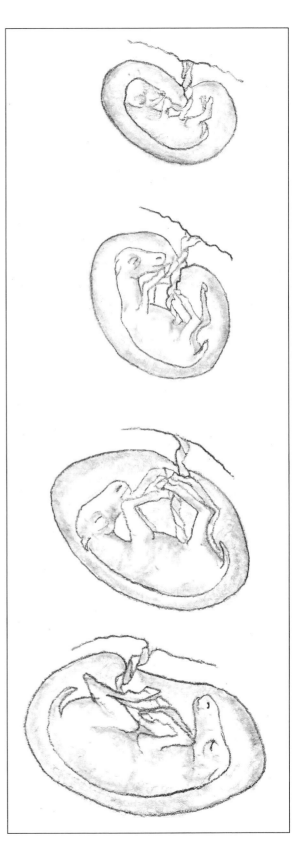

The embryonic foal first develops those features and systems which will enable it to survive life, and to grow in the womb; those which enable it to survive in the outside world develop later in the pregnancy.

Nutrition is, of course, vital to the life of the embryo, as is a good system of gaseous exchange – the receipt of oxygen and the disposal of the potentially toxic waste products of metabolism (carbon dioxide, ammonia and others). For these purposes, the embryo, later termed a foetus, develops inside two highly specialised membranes, the placenta (the chorio-amniotic membrane), which lies next to the endometrium, and the amnion, which directly surrounds the foetus. Contact with the foetus is through the umbilical cord and through the placenta itself. The cord carries vital blood to the foetus containing nutrients, gases, hormones and other substances, but the mare's and foal's blood supplies do not actually mix. The cord also carries away foetal urine.

By the hundredth day of gestation, the placenta is securely attached to the entire wall of the uterus by millions of tiny projections or villi containing blood capillaries, by means of which exchange of substances can take place, in addition to the umbilical cord.

Between the placenta and the amnion is a 'cushion' of allantoic fluid. This largely comprises waste products composed partly of foetal urine and partly of products formed by the placenta. The foetal urine is discharged into this area through a special duct in the umbilical cord called the urachus, which closes when the cord breaks after birth. The allantoic fluid protects the foetus to a large extent from physical impact (broodmares being only a little less prone to frolicking about, rolling, lying down and playing than other horses). This is the fluid which escapes when the mare's 'waters break' during the first stage of parturition or foaling.

The allantoic fluid contains a pad of rubbery tissue, the hippomane (which means 'horse madness') which was formerly thought to have special magical properties, possibly because no one knew what it was. Old-time 'horse whisperers' used to carry dried hippomanes around with them and it is possible that some horses were intrigued by the smell and became diverted and thus less aggressive and more co-operative towards this horsey-smelling human. I have known this practice to continue well into living memory, and it may work. We now know that the hippomane simply consists of waste body debris, cells and salts. To my knowledge, no one has proved that it possesses any magical powers!

The amnion is a white membrane or sac surrounding the foetus. Inside it, the foetus is bathed in its own fluid — amniotic fluid — containing some foetal urine plus its own slimy secretions which acts as a protective lubricant during the foal's movements in the uterus. The amnion usually breaks naturally from around the muzzle during foaling but sometimes the foaling attendant needs to break it by hand (which is easily done), so that the foal can start breathing without delay.

In this highly specialised, protected environment, the embryo develops into a foetus and the foetus, we hope, into a healthy, full-term foal. Interesting opinions have been expressed about whether the foetus is already aware of its

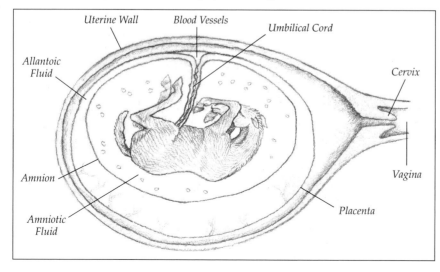

Diagramatic representation of the foetus in the womb. The foetus is shown much smaller than it would be at this stage (note the hair coat present) for clarity.

dam's smell and taste and the sound of her voice before birth. This seems reasonable, but it is known that newborns of higher species psychologically attach themselves to, or become 'imprinted' upon, almost any object (even inanimate ones) which they see very shortly after birth, whether they are already familiar with them or not. This subject will be dealt with in the next chapter.

The foetus moves around somewhat during development and appears to people observing the mare to be kicking out and generally moving about. Most of the structures are well developed long before birth. The hair coat, for instance, develops well beforehand, and the heart and circulation are also in full working order, if at a slower pace than after birth. The muscles, kidneys, liver and nervous systems are in place and may be working, and the systems for the production of hormones, enzymes and other body chemicals must be well established if the newborn foal is to survive. The lungs are fully developed but are obviously not working in the way they will soon after birth.

The movements of the foetus in the uterus help prepare the foal for the sudden surge in the demands made on its body after birth when, although it has its dam for external protection and initial nutrition, it will largely have to fend for itself physiologically.

Hormonal control of pregnancy

Hormones control both the maternal and the foetal inputs to pregnancy and work together in co-operative, balanced and extremely complicated processes, some being responsible for the mare's side of things and some for that of the foetus.

The prime hormone maintaining pregnancy is progesterone, known as the pregnancy hormone. As well as being responsible for the states of dioestrus and anoestrus, as explained in chapter 2, progesterone is responsible for promoting the favourable state of the uterus to receive the zygote. It is also needed for the zygote's implantation into the uterine wall and for the continuing attachment of the placenta (by which progesterone is secreted) to the uterine wall.

The role of other hormones, however, is vital to the whole process and research in recent years has greatly increased the veterinary profession's knowledge and understanding of this intricate process. Nevertheless, it is still far from fully understood.

Pregnancy failure

Although the average gestation lasts 330 to 345 days, foals born after 320 days are regarded as full-term, but they may be weak (dysmature). Those born from 300 to 320 days are premature and not fully developed; they may well survive with help. Those born before 300 days however, cannot survive, they are simply not well enough developed to withstand life outside the womb. The range between which foals should survive is from 310 to 374 days.

Mares may abort before 320 days for various reasons, infectious and non-infectious (foals born dead after 320 days are termed still-births). Infectious abortion may be due to bacterial, viral or fungal infections, which may result in a diseased placenta which is unable to serve the foetus. Infections may also affect the foetus itself to the extent that it cannot develop or survive and the mare's body rejects and aborts it. And although mares with diseased genital tracts can conceive, their pregnancies often seem to terminate early – another reason for checking the health status of the mare before she goes to stud.

She should receive whatever vaccinations are currenty recommended – take your vet's advice. Although, as far as general health is concerned, some experts are beginning to feel that conventional vaccination can actually undermine the immune system rather than assist or enhance it, few owners are likely to want to take the risk of not having their mares vaccinated against any disease which may adversely affect their health or pregnancies.

Homoeopathic vets are naturally in favour of homoeopathic vaccinations, and this form of medicine and healing is certainly becoming increasingly popular. Whatever your feelings on the subject, it is one which you should not neglect to discuss fully with your vet.

The presence of twins in the uterus usually results in an abortion, if it is not spotted and terminated very early by the vet, because the fact that the placenta is attached to the whole of the uterine wall means that there is not usually room for a second foetus and placenta. Occasionally twins are born and reared successfully but they are generally smaller and less robust than single foals. Usually, one is dominant in the womb and deprives the other of essential supplies, so the latter dies and is aborted. The survivor very often follows, as a result of hormonal insufficiencies or underdevelopment which cannot be made up. If your mare conceives twins, you can either have both aborted or just

37

one. Again, discuss this with your vet.

Hormonal abnormalities are likely to be a cause of abortion if there is no evidence of disease. Because of our incomplete knowledge of hormonal control of pregnancy, there seems at present to be little that can be done to maintain the process, although in cases where hormonal failure may be anticipated, perhaps detected by blood tests, injections of progesterone have been used. This method is not regarded as very satisfactory, however, as there is little evidence that it actually works. At present, it seems that little can be done to detect or prevent abortion due to hormonal abnormalities.

Another reason for abortion is a twisted umbilical cord. Because the foetus moves around in the uterus, it is possible for the umbilical cord to become misplaced, especially if it is too long. It may then either knot or twist itself, so cutting off its channels of support, or become wrapped tightly around the foetus, cutting off the blood circulation from the placenta. Sometimes, it does not become misplaced until birth, when the foetus moves around dramatically to prepare itself for the correct positioning and presentation. It may then become twisted around the neck, causing a still-birth.

Sometimes, perhaps most often for genetic reasons, the foetus may simply not form and develop normally, and may become deformed to the extent that it cannot maintain life even in the womb, let alone outside it. It seems that old mares are more likely to produce deformed foetuses or foals, perhaps another good reason for not continuing to breed from them.

Trauma, distress, shock and malnutrition may cause abortion as a result of changes in the body's biochemistry. An accident in the field or yard could be responsible, such as the mare falling, becoming cast or entangled, galloping into something or being seriously kicked. A traffic accident would obviously be extremely traumatic, as would an attack by dogs or other animals. Distress may result from being separated from a close friend or other horses, rough treatment from handlers, unpleasant veterinary treatment, some drugs and a bad attack of colic. Indeed, anything stressful in the vicinity (such as low-flying aircraft which are believed, from incidents in the USA, to cause abortion) may cause the mare's body chemistry to change to the disadvantage of the foetus.

It is hard to pinpoint hard evidence for these causes of abortion but it seems logical that they could be a cause, and it would be sensible to try to arrange as relaxed and uneventful a pregnancy as possible. Good preventive management would consist of giving the mare a correct, balanced diet, ample clean water, congenial companions, good shelter, safe housing, plenty of liberty, gentle exercise and a calm, relaxed lifestyle. Arrange a good routine in a familiar environment with people and animals she knows and trusts.

One obvious sign of abortion is the mare standing over her foetus, but a mare which refuses to leave the field when the handler tries to bring her in at a normal time may be reluctant to leave an aborted foal. This is not so obvious and may only be discovered when the handler goes back to investigate. Moreover, if the mare's temperament returns to normal, if she fails to increase in size as expected, if she does not become slightly more lethargic as time goes on and, of course, if she comes into season, you can strongly suspect that she has aborted without your knowledge. And if, at any time, you notice any unusual discharge or blood coming from her vulva, you should certainly call the vet at once. You should also call the vet if she shows colic-like behaviour, signs of pain or discomfort or general unease.

Should you be unfortunate enough to have your mare abort, she should immediately be isolated, if at all possible, until the cause is known. Various veterinary checks can be done and, if you find the foetus or associated membranes, they must be kept for the vet to take away and examine. Any area with which she has been in contact should be thoroughly disinfected, her bedding burnt and normal disease and isolation precautions taken. If an infectious cause is ruled out, you can return to normal.

Pregnant mares which are in contact with one which has aborted are at great risk, of course, so abortion should not be regarded merely as a sad (or inconvenient and expensive) event but as a serious health risk to the entire premises. Some diseases can be passed on from and to non-breeding horses, so general good hygiene and an effective preventive medicine regime are important in all stable yards and studs. Remember that infections can persist in animals, sometimes undetected, for months or even years. Premises can be contaminated likewise so, inconvenient though it may be, good management is not enough. You need the back-up of good hygiene and sensible health practice to be as sure as you reasonably can that neither disease nor poor management practices will thwart your breeding efforts.

Chapter 6
Before, During and After Foaling

Probably over 95 per cent of foalings go off without a hitch and often without human intervention or presence. So novice breeders need not be put off by what they have read so far. The more practical aspects of foaling are dealt with in chapter 17; here we will consider veterinary matters and what will happen during a normal run-up to foaling, the foaling itself and the aftermath. Difficulties and abnormalities are dealt with in chapter 9.

Countdown to foaling

Up to about a month before the expected birth date, the mare will still be reasonably active. But the foetus will have been making heavy nutritional and metabolic demands on her for three or four months and will continue to do so for a similar length of time during lactation. An ample, balanced, nutritious diet is therefore particularly vital to broodmares for three months before and three months after foaling. This does not apply to ponies and cobs, however, who do not require extra nutrients until foaling.

At this point (a month before the expected birth date), the mare will begin to show physical signs of preparing for foaling. The muscles, tendons and ligaments of the hindquarters begin to soften to facilitate birth and she may appear to sag in this area. The birth canal itself begins to

---

NEONATAL ASSESSMENT AND SCORING OF FOALS

Normal birthweight of thoroughbred foals: 42–46 kg (Primipara), 48–52 kg (Multipara)

NORMAL FOAL ASSESSMENT AT BIRTH

TIME	Temperature (°C)	Pulse (/min)	Resp. Rate (/min)	Notes
0–1 min	37–37.5	70	70	Hypoxia/metab + resp acidosis
5-30 min	36.8–37.0	120	50	cord rupture/shivering/righting, attempts to stand/sucking reflex
30–60 min	37.5–38.2	140	40	Co-ordination/standing/Maternal recognition
1–2 hr	38.0	120	35	Teat seeking, follows mare
2–12 hr	38.0	100	35	Meconium/urine passed/COLOSTRUM ESSENTIAL
12–48 hr	38	90	30	Bonding. Closure of Foramen Ovale + Ductus Arteriosus NO more colostral absorption

BIRTH SCORE (assessed in first 5 minutes of life)

SCORE	0	1	2
HEART RATE	Undetectable	< 60	> 60
RESPIRATORY RATE	Undetectable	Slow/Irregular	Regular > 60
MUSCLE TONE	LIMP	Flexed extremities	Sternal recumbency
NASAL RESPONSE	NIL	Grimace/Movement	Sneeze/Rejection
INTERPRETATION	Normal foals score 7–8 Moderate depression 4–6		Marked depression 1–4 DEAD 0

Hypoxia: diminished availability of oxygen to body tissues. Multipara: a mare which has had two or more pregnancies resulting in viable offspring. Primipara: a mare which has had one pregnancy which resulted in viable offspring.
(*Reproduced from* Formulary of Equine Medicine *by D.C. Knottenbelt, published by Liverpool University Press.*)

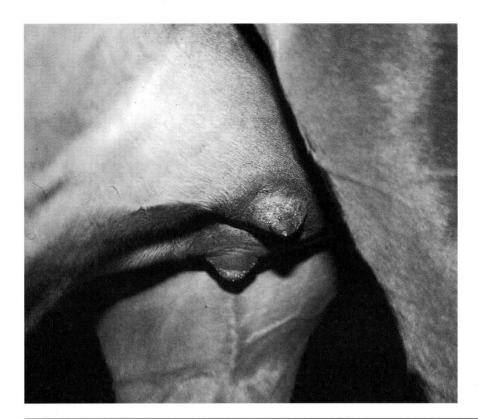

Some time before foaling, the mare's udder will increase in size (called 'bagging up') and a waxy deposit may form on the teats. Unfortunately, this does not always mean that foaling will occur within the next day or two. (Vanessa Britton).

loosen up, and the uterine muscles start to work and give the mare twinges of discomfort as they tone up for the very hard work they will have to do during foaling.

The mammary glands or udder also start to increase in size, but without producing milk. As the expected date draws nearer, you can, if you wish to go to these lengths, have a small amount of first milk or colostrum analysed, perhaps daily. By assessing the changing levels of sodium, calcium and potassium salts, your vet can gauge the probable time of foaling to within twenty-four hours.

The foetus, although fairly active, will be lying on its back most of the time. All its major organs and body systems will be developed, but they still need to mature somewhat. Its skeletal system is still more cartilage than bone, although by the time of foaling it will be half and half. The body tissues, especially those which will be exposed to the outside world, toughen up and the nervous system rapidly matures.

The foetus now moves around more (while it has the space – it is still increasing in size) and appears to start trying to move and stretch its limbs and exercise its muscles and joints. Although its food supply has come from its blood and not by mouth, at this point it begins to swallow the amniotic fluid surrounding it to gear up its intestines. The fluid contains some remnants of nutrients which the intestines can use. The final waste products accumulate at the lower end of the intestinal tract to be voided as the foal's first dung or meconium, probably just a few hours after birth.

About a week before birth, hormonal secretions from the foetus act on the mare's own hormones, reducing their effect and warning the mare that it is becoming ready to be born. With little room for manoeuvre now, the foetus nevertheless moves into a more upright position in the uterus, although still mainly on its back. This gives the mare a different external shape, her belly now seeming to hang even lower. This has a pulling-down effect on her soft tissues and she may even appear to be losing weight over her back and loins.

Many mares will start to 'wax up' at this point, accumulating on the ends of the teats a honey-coloured, waxy substance consisting of antibodies and serum secreted by the udder. Those mares which actually start to run milk at this time, losing vital colostrum, have usually been overfed.

The mare will probably be quite uncomfortable now; the uterus will be tensing up and relaxing in preparation for foaling contractions. She may become slightly constipated due to the internal pressure on the lower part of her intestinal tract. Her legs may fill because of lack of exercise and decreased circulation, owing partly to internal pressure and carrying the extra weight of her foal. The cervix, vagina and vulva now start to soften and relax to assist birth. The mare's behaviour will probably show her increasing discomfort. She may become slightly irritable or nervous, urinate less but more frequently, have some discomfort when passing droppings and sweat-up slightly. Some mares perform the flehmen action quite often, presumably as an expression of their discomfort and possible frustration. Some become reluctant to lie down because it is difficult to get up again, but some lie down more than usual.

A few mares may experience actual digestive problems owing to the pressure on the internal organs. Check your mare's droppings and urine and the way she passes them (or fails to) as often as you see her attempting to defecate or urinate. Note her behaviour and the normality or otherwise of her droppings and urine, and if you are uneasy about anything consult your vet.

One day before birth, the hormonal signals from foetus to mare are stronger and the foetus works hard to get into the correct position for birth. It twists, at least partially, onto its belly, with the forelegs extended and the head resting on the knees, the hindlegs still being tucked underneath the abdomen. At the same time, the uterus is tightening around the foetus, making this change of position against powerful resistance quite a muscular effort – good practice for the future. Its front hooves and muzzle may be inside the pelvic canal, helping to stretch the area in readiness for birth.

The mare will have 'bagged up' now and her udder may be hot, tight and tender, possibly with some milk leaking out. Her vulva will be considerably enlarged and the area of the flanks and the root of the tail will be loose and sunken. The cervix will be partially opened. In her behaviour, the mare will start to become independent, looking for private areas to foal and not wanting to bother with other animals or people. If her attendants have not yet started to prepare for sitting up with her they should do so now or they might miss the whole event!

The complex roles of hormones have already been mentioned several times, and when foaling

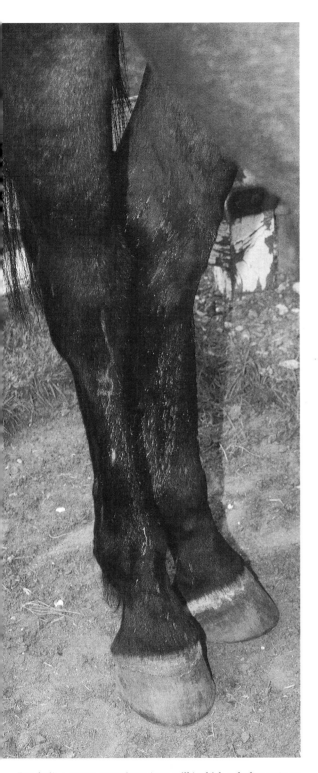

Pre-foaling mares sometimes 'run milk' which splashes onto or runs down the insides of their hindlegs. This often happens in mares who have been very well, or overfed. Should the process continue for some days, vital colostrum ('first milk') will be lost. (Vanessa Britton).

is imminent, they continue to play their part. The foetus has been living a life of its own in the uterus, albeit dependent on the mare for supplies and maintenance, and it is the foetus itself which starts sending out hormonal messages to the mare to announce when it is ready to be born. The mare receives these messages but it is she, not the foetus, that decides the actual time of foaling.

As a result of millions of years of conditioning by evolution, most mares prefer to foal in darkness and privacy. In the early stages, they are even capable of postponing their labour if they feel they are being disturbed. They will prefer to stay apart from the herd if they are part of one, and in domestic conditions, although they may like to know other horses are around, they normally do not mind being moved to a more private and bigger box for foaling, or a nearby paddock which may be handier for their attendants if they are foaling outdoors. It is thought that the hormones which initiate birth are stimulated by decreasing light levels in the late afternoon and evening, and inhibited as light levels rise towards and just after dawn, which accounts for the habit of foaling in darkness.

The hormonal interactions which trigger and control the foaling process are still being studied and are not yet completely understood. But it appears that the hormones cortisone and reducing oestrogen (plus a rise in progesterone which triggers lactation) prepare the muscles of the uterine wall for instructions from oxytocin, prostaglandins, relaxin and corticosteroids to begin the actual birth process, starting the contractions of the uterine muscles which gradually push the foetus out.

It may be tempting for busy, working owners or even commercial studs with a heavy workload to consider inducing foaling by injecting oxytocin or prostaglandins, at least in mares very near or apparently over their time, in order to facilitate everyone's life and work schedules. This is probably not a good idea, however, unless the vet really feels the mare should be induced for her own wellbeing and that of her foal. Unless they are obviously in difficulties, induction can result in the foal's being born earlier than is best. The result will be a foal which in his terms is premature, having been denied a little extra time for maturation – not a good start to life.

The stages of labour and parturition

Labour and the foaling process are divided into three stages called, logically, first-, second- and third-stage labour. First-stage labour is the period before the placenta breaks and the allantoic fluid either gushes or trickles out of the vulva – the 'breaking of the waters'. Second-stage labour starts at that point and is the actual delivery of the foal, and third-stage labour is the explusion of the placenta, amnion and umbilical cord as the afterbirth. The actual foaling process can take from as little as fifteen minutes to an hour – most foalings take less than an hour.

About an hour before birth, the muscles of the uterus contract firmly and regularly from front to back in a wave-like movement, with short breaks in between for them to recoup their energy before tightening again around the foetus, gradually pushing and guiding it forward towards the birth canal. The mare will probably be sweating somewhat, with her surface veins prominent, and she may possibly be running some milk from her udder. She will have been uneasy, uncomfortable, nervous and showing colicky signs, walking the box or paddock, pawing the ground and shifting her weight from foot to foot, looking round at her flanks, repeatedly lifting a hindleg and probably kicking, swishing her tail, performing the flehmen posture and possibly groaning slightly. She may stand pressing her tail against something firm and solid such as a wall or tree and will also probably pass small amounts of faeces and urine quite frequently in an effort to relieve the pressure inside her.

She will almost always be lying down as the

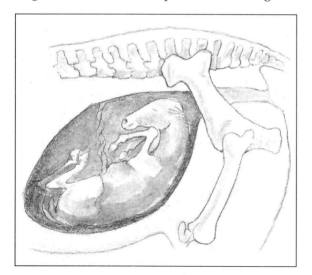

During the first-stage labour, the foetus is initially on its back.

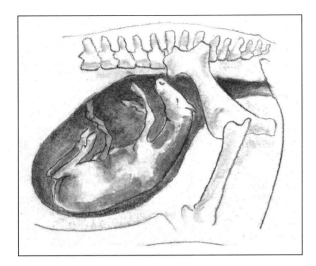

It starts to twist round ...

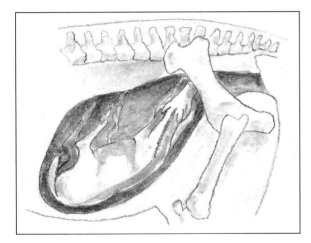

... and extend the forelegs ...

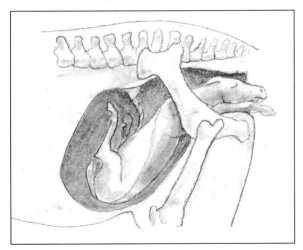

... until the forelegs and the head are extended through the pelvic ring and birth canal.

contractions continue and it is the pressure of her abdomen against the ground, combined with the now serious lack of space, which is actually the main explosive force in this process.

First-stage labour can be frustrating for the mare, her handlers and, presumably, the foetus because it can last anything from a few minutes to a few weeks and may come and go, confusing everybody, At this point, novice breeders may well be very worried and seek veterinary assurance that this is not unusual and that their mare is all right.

The foetus at this stage is twisting itself from the front end, from lying on its back to gradually lying on its belly and chest with its forelegs extended into the birth canal, the head resting on the knees. There is no going back now. Second-stage labour can last from five to twenty-five minutes, and averages about fifteen. If within half an hour the waters have broken and the mare seems to be making no progress with foaling despite trying (and the foal's two forefeet are not showing through the vulva), it is advisable to call for veterinary help. A few mares will not show signs of first-stage labour but foal very quickly in second-stage labour.

The cervix is now fully open and the mare will be straining and pushing with her abdominal muscles with each contraction, supported by the diaphragm muscle (which separates the chest from the abdominal cavity and is instrumental in breathing). The diaphragm muscle will be firm and static as she holds her breath for extra force. She will also be supported by her spine which she may, in turn, support by lying with her back against the wall if she is foaling indoors.

The outermost of the foetal membranes, the placenta, will be broken by the foal's front hooves pawing and pushing on it just inside the cervix. This causes the allantoic fluid to be released into the birth canal, lubricating it and giving the first unmistakable external sign that foaling has begun, although novice breeders may confuse this with the mare urinating. The voiding of this dark yellow-coloured fluid, up to about 9 litres (2 gallons) of it, releases some of the pressure the mare and foetus have been feeling. The former may now stop sweating, stand up and even nibble some food and have a drink. She will probably smell, lick and nibble the straw wetted by the allantoic fluid. This helps with the bonding process between her and her foal and even keeps her with her newborn until it can stand and follow her, rather than wandering off to graze without it and leaving it in danger.

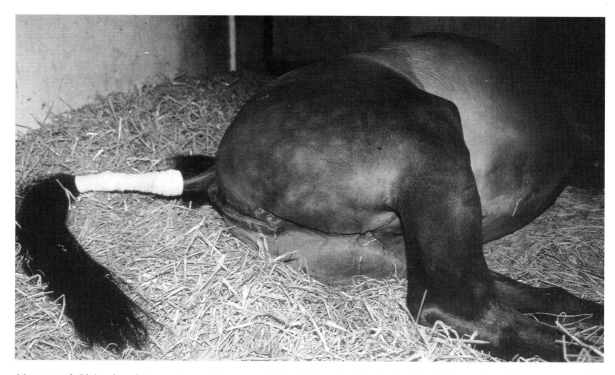

Most mares foal lying down but may get up and down several times before actually beginning to foal. The muscles and ligaments around the tail will soften and the vulva will enlarge, as here. (Vanessa Britton)

The foal's forefeet appear at the vulva, one in front of the other and still in the fluid-filled sac. (Vanessa Britton)

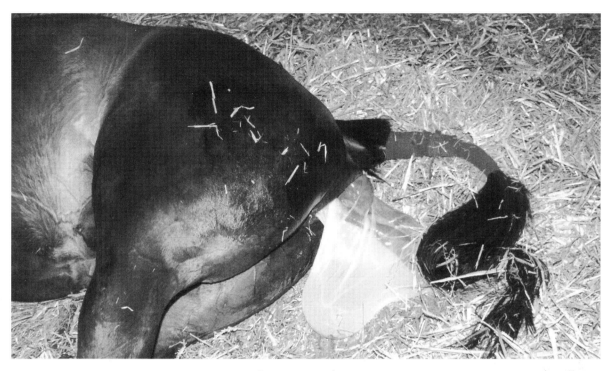

The muzzle appears – on top of the forelegs as it should be. (Vanessa Britton)

If the mare is having trouble expelling the foal's shoulders, which are its bulkiest part, an experienced stud assistant or the vet can gently but firmly pull on the forelegs in a downward direction (towards the mare's hocks) in time with her contractions, keeping one leg in front of the other to angle the shoulders and make it easier for them to pass through the bony pelvic ring. (Vanessa Britton)

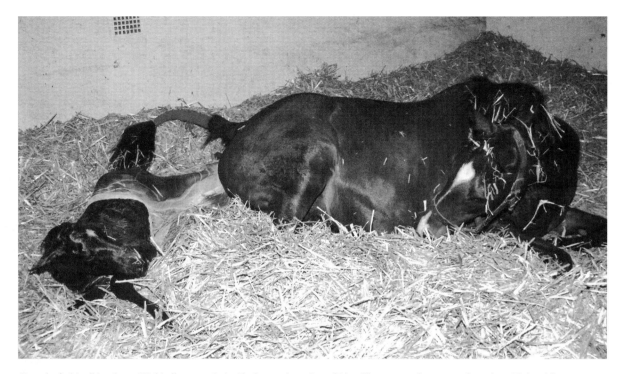

Here the foal is all but born. His hindlegs remain inside the mare's vagina which will encourage her to stay down (provided nothing or no one disturbs her) as they both rest and blood continues to pump down the umbilical cord into the foal. The sac may split on its own but if it does not, tear it manually as soon as possible to free the foal's muzzle. Clear the nostrils of fluid and any debris and check that he is breathing. (Vanessa Britton)

The mare has risen and the umbilical cord has broken. She licks and nuzzles her foal to dry him and familiarise herself with his taste and smell. (Vanessa Britton)

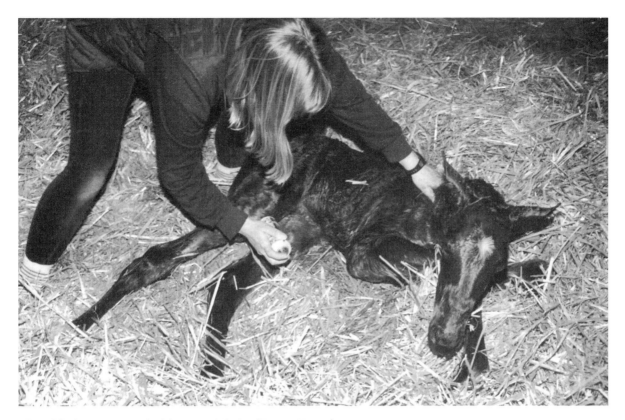

The umbilical stump is treated to help prevent infection. (Vanessa Britton)

She may lie down and get up several times during second-stage labour, but horses, unlike some other species, rarely foal standing up. It is more likely that the rising and lying down, the slight attempts at rolling, the stretching and the general moving around may help the foetus to get itself into the correct position for birth, or that the mare herself is helping to position it.

Soon the pain starts again as stronger contractions recommence, occurring in waves as described, from the front of the uterus to the back to squeeze the foal strongly towards the birth canal. The mare will probably lie flat on her side, adding as much ground surface as possible to the expulsive force of the uterine muscles; her instinctive impulse to push carries on until the foetus's forelegs, head, shoulders and chest are through the bony pelvic rim.

The amnion will probably first show protruding through the vulva as a fluid-filled balloon-like membrane, pearly-white to bluish-grey in colour, through which the two front hooves can be discerned. An experienced stud hand may carefully put a hand inside the vagina now to check that the foal is in the correct position and presentation for foaling, to avoid undue straining by the mare trying to deliver a foal with, say, the head or a leg bent backwards, so blocking its own passage.

The foal should come through with one foreleg slightly in front of the other and the head resting on the knees. In this position the elbows and shoulders are angled one slightly in front of the other, and so form a narrower profile to be pushed through the pelvic canal. This part of the foaling process is a major hurdle to be overcome.

As first one front hoof then the other appear through the vulva, they may still be covered by the amnion or not, and it may soon break anyway, as the foetus continues to be pushed out. At this point the foal, as we may now call it, will lie partly outside and partly still inside the birth canal as the mare briefly rests again.

If the amnion is still covering the muzzle, the attendant may break it with a hand and carefully clear the foal's nostrils of mucus and any debris so that it can start breathing without delay. The tremendous pressure of being forced through the pelvic canal, unlike anything the foal has ever experienced before, will have squeezed the

47

ribcage. This will have strongly stimulated the chest muscles and the lungs to start working, clearing out the fluid that has been in the lungs and taking in air. The foal can then take the first essential breath to start its new life outside the womb, where breathing was neither desirable, necessary nor possible. As the chest and abdomen are pushed through the canal, the umbilical cord can be squashed against the bony pelvis, which greatly decreases the blood (and therefore oxygen) supply coursing through to the foal. It is possible that this, too, may act as a stimulus to the foal to start the breathing process. It will gasp and splutter for a while as its lungs gradually 'learn' to expand and the chest muscles respond to the nervous stimuli which make them work. It is essential that the foal starts breathing as quickly as possible once its head is through the vulva, as oxygen deprivation can result in damage to the brain and nervous system.

Very soon after the foal starts to breathe, a special short blood vessel (or hole) between the left and right sides of the heart, called the ductus arteriosus, which has linked the pulmonary artery to the aorta, begins to close and will no longer be functional after about four days. Its purpose has been to pass blood through the heart's dividing wall rather than its being sent through the lungs as in a mature horse, because the lungs are not functional before birth. Its job done, it gradually closes up after birth. Until it does, it may give the impression that the foal has a heart murmur.

When the foal is first born, its tongue and gums may seem a pale, even bluish-white colour, owing to slight oxygen deprivation and the build-up of carbon dioxide during birth. However, once its respiratory system gets under way they will very quickly become a deeper pink than the normal salmon shade of an older equine.

The mare may turn up onto her breastbone to recuperate somewhat after the exhaustive physical and mental work and stress of pushing out the foal's forehand. She will probably turn her head and neck to inspect and smell her new half-arrival. This is one of the most magical and appealing parts of the whole process, certainly to the human attendants, and probably to the mare herself.

A few final contractions will push out the hindquarters and the hindlegs may well stay inside the mare, probably encouraging her to stay lying down and so giving the pair a chance to rest. The foal, however, may struggle and kick its legs free. It is important not to disturb the mare now but to let her stay more or less as long as she wants to (unless the time is becoming excessive – say over an hour) as it is best to let the umbilical cord break naturally in its own time. It will seal more effectively, helping to prevent bleeding and the entry of disease organisms. Blood is still passing to the foal through the cord at this point, so breaking the cord early will mean depriving the foal of essential blood.

Up to a couple of generations ago, the cord was routinely cut at birth, but it was later thought that early breaking of the cord would deprive the foal of up to a third of his own blood volume, as blood still flowed into him down the cord. More recent research has indicated, however, that the transfer of blood down the cord stops very quickly after birth. Nevertheless, most vets and experienced attendants still feel it is generally best to leave it alone for a while. Although the conflicting views on whether or not blood does still transfer to the foal down the umbilical cord is confusing, your vet will be able to give you the latest information on this matter. Whatever the situation, the general advice is to let the cord break naturally whenever possible.

Whilst the pair are still lying down (the mare will normally stay down for about forty-five minutes), the mare usually starts licking and nuzzling the foal, and may whicker to it. This all helps mutual recognition and bonding and also the stimulation of the foal's skin and muscles to warm it up and dry it off. Most mares are very interested in their foals. Bad broodmares and some maiden mares may be thoroughly distressed and frightened by the whole process and may take an instant dislike to the foal, treating it as an intruder and maybe even trying to savage it, particularly when it first tries to suckle. However most bond quickly and soon become very protective towards the foal. They vary in their attitudes to human intervention.

The umbilical cord will normally become slightly thickened and pale at a point a very few centimetres from the foal's navel, and will break naturally when the mare stands up and exerts a pulling force on it, or when the foal starts to try to stand up and does likewise. If it does not break, firm but careful pulling on either side of this point should bring it apart. In this case, the attendants should try to ensure that any undue bleeding from the stump is stopped by tying the end fairly tightly with sterile tape or a sterile

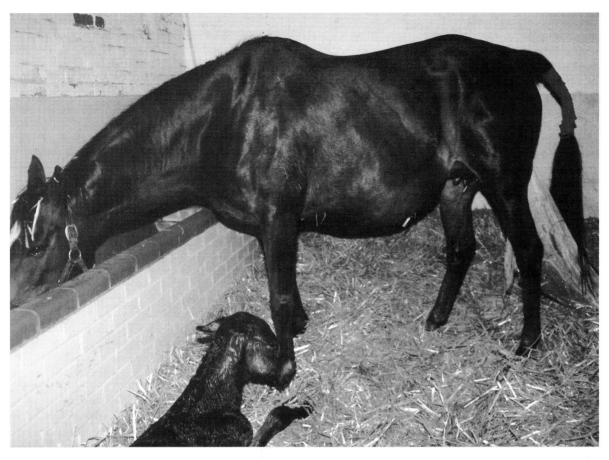

The afterbirth has started to come away and could now be tied up to prevent the mare treading on it, pulling it out and tearing herself internally. She enjoys a much needed feed whilst the foal remains down, resting after one of the most stressful experiences of his life. (Vanessa Britton)

bandage (which is kept wrapped until needed). This should not be necessary if the cord has been allowed to break in its own time. The stump is dipped into or sprayed with a disinfectant recommended by the vet to help prevent disease entering the foal's body.

The final stage of parturition is the expulsion of the placenta and amnion, together with the umbilical cord, collectively called the afterbirth. This process is called 'cleansing'. After the birth, the uterus rests a little but when the foal first starts to suckle, although often before this, the hormone oxytocin secreted by the pituitary gland stimulates more uterine contractions and also the let-down of milk, the protective, energy-giving and laxative first milk or colostrum. Fresh uterine contractions to expel the afterbirth will usually begin again once the placenta has started to 'peel' itself away from the uterine lining. The mare experiences lower-level pains of contraction, like those of birth itself, during this process so may lie down again while it is going on.

The afterbirth may hang down around the mare's hocks or even lower for some hours and, if she or the foal treads on it the uterus could be injured by its being torn away prematurely. Any injury to the inside of the uterus favours the entry and development of disease so it is normal practice to tie up any membranes which do not come away immediately with binder twine or string, to prevent this. This also creates a light weight which may assist in cleansing, but it is certainly not advised that you pull at the membranes – other than perhaps a light hand-shake type of tension in case they are simply lying loose in the birth canal – as you could either leave part of them inside the mare or tear and injure the uterus itself. It is advisable to seek veterinary advice if the afterbirth is not expelled completely within about three hours of birth, certainly within six hours, as its retention can also cause serious disease.

The afterbirth should be kept in a plastic bag in a cold, dark place and given to the vet to examine, whether or not you suspect problems, as it is essential that all the placenta comes away. Any portions of membrane which are retained within the uterus are likely to start to decay and cause disease. You can check the membranes yourself, of course, and if there are any obvious holes and missing pieces tell the vet at once, but novice breeders, particularly, should save the afterbirth and give it to the vet anyway, for safety's sake. Be aware that a few mares try to eat the afterbirth (even though they are herbivores) which may be an ancestral reaction aimed at reducing smells attractive to predators and obtaining some immediate nourishment to replenish used energy. It is safest to prevent this so that you can check that it is all present.

The expulsion of the afterbirth completes the process of labour and parturition and you will by now, it is hoped, have a healthy, if very tired, mare and foal.

After the birth

The foal's breathing and heart rate will initially be fast and irregular. It has to operate at a higher general level than when it was inside its dam and this is hard work and takes time to establish – up to twenty-four hours. Within this time, the respiratory rate should settle at thirty breaths per minute (in and out counting as one), the at-rest heart rate at about ninety beats per minute and the temperature at about 38°C or 101.5°F. It will be 37–37.5°C or 99–100°F at birth.

The foal will have been born into a world considerably colder than the womb, and its temperature regulation mechanisms need time to adapt. The foal-coat will dry off and become fluffy and insulating but not yet capable of resisting and protecting it against wet weather. Until it starts to dry off and warm up, aided by the dam, it will probably shiver quite a bit. This exercises the muscles as well as helping the foal warm up.

Because the instinct to be up and able to run with the herd is strong even in domesticated foals, it will not take too long to struggle to its feet and stand up, if it is healthy and normally alert. It will be propping itself on its sternum or breastbone within about ten minutes to half an hour, which will help it breathe better, and will exert considerable muscular effort in trying to hold its head up and steady. This is its first full experience of the ultimately irresistible force of gravity, against which it will be struggling for the rest of its life – like all of us.

It will start trying to control its gangly legs against the new environment of gravity, ground and lots of space and air within half an hour (often less) to two hours at the most. Speed is of the essence because, within a couple of hours, it should take its first drink of colostrum, which contains antibodies against disease (against which it has no natural protection of its own yet), glucose for energy and a laxative substance to help with the first dropping, getting rid of the meconium which has built up in the hind part of the intestinal tract.

Within about an hour, or up to two hours, then, the foal should be up on its feet, balancing precariously and falling down again several times. Getting onto its feet is obviously very hard work. As it tries to balance, its own intuition tells it that by moving a hoof this way or that it can avoid falling over and, in this way, it starts to take its first steps, forwards, sideways or backwards. The weight of the head being pulled ever downwards towards the ground by gravity might suggest that forwards is probably a good way to go and the forward impulse (essential to horses for their survival) is confirmed. Feral and wild equidae are able to run, if a little uncertainly, within three or four hours of birth and so are domestic foals if given the space to practise.

All the hard work involved in being born and standing up has probably made the foal feel hungry, a sensation it will not have experienced before, and this feeling will tell it that it needs to find a new way of acquiring nourishment. Unfortunately, it will have no idea what the dam's udder is for. Indeed, it will still not be sure what *she* is and, to make matters even more confusing, will not yet be able to see as well as an adult horse, even though it was born with its eyes open.

It will stumble and fumble around, looking for what it will later recognise as the dam's udder and a source of food. Experienced mares will position themselves to make this easy for their foals, relaxing the hind leg away from them to tilt the udder towards them. They will push, pull and guide the foal's search with their muzzles until it finally latches onto one teat. It may already have previously tried an elbow, a hock, an attendant or even some inanimate object into which it has bumped and which satisfies its instinctive desire to find something small and suitable for sucking, above head-height and in a dark-ish, enclosed space. The teats meet all these requirements, of course, and once found they are never forgotten! The young, healthy foal will suckle enthusiastically for a minute or so several times an hour, suckling for longer but less frequently over the days and weeks each time it feels hungry.

As a foal has no immunity to disease when born, relying entirely on the antibodies in the dam's colostrum for protection, it is essential that it drinks as soon as possible, ideally within six hours. The gut is capable of absorbing antibodies for a maximum of forty-eight hours, although the ability to do so decreases from six hours of age onwards, so the importance of learning to suckle as soon as possible is obvious.

Once the colostrum has been absorbed, the feeling of fullness, the natural laxative in the colostrum and the energy-giving glucose will encourage the foal to perform another new action, that of doing a dropping, the first being the rubbery, dark-coloured meconium. Soon the droppings will become mustardy in colour as its exclusively milk diet works its way through its system. Pretty soon, too, it will be stimulated to stale or pass urine, and will learn the two new stances these actions demand. Falling over whilst learning them is par for the course.

The suckling instinct is one that the foal uses not only for obtaining nourishment, but also as a comfort reaction to anything that upsets it and promotes the need for bonding with and protection from the dam. If a human or any strange animal approaches, for instance, it is normal for the foal to get up and go behind the mare for a drink. A sudden noise will have the same effect, as will the attentions of other animals and so on. In fact, if this reaction is not noted, it is possible that it is not feeling well or may have some sort of dysfunction.

Significant changes occur in the foal's blood chemistry during the first twelve or so hours of its life, and your vet can have those monitored from a blood sample if this is felt to be desirable. Apart from checking the concentrations of various nutrients and chemicals, the levels of immunoglobulins (antibodies against disease gained from the mare's colostrum) can be checked: they should rise steadily up to forty-eight hours of age, and if they do not your vet can supplement them intravenously.

It is said that a newborn foal achieves in its

first hour of life what it takes a human baby a year to accomplish. This is essential and a remnant of life in the wild which is still strongly present in domesticated horses and ponies.

The mare has obviously undergone a good deal of pain, physical exertion and mental stress during the process of giving birth. She may have experienced sharp pains in her udder when the early process of milk production and let-down occurred. She may also have found that the foal's suckling caused soreness or an unpleasantly sensitive feeling, although both of these usually improve fairly soon. Then, just when she thought it was all over, the afterbirth pains started and she had to go through a mini-birth process again to expel the placenta. Physical signs of all this may be a slightly raised temperature from the effort and increased heart and respiration rates, all of which should return to normal in well under an hour. The udder's tightness and soreness gradually lessen, as does any swelling of her belly and filling in her legs.

Her behaviour towards her foal, if she is a good mother, will become very protective and she will be interested in little else. This can call for firm but tactful handling from the stud staff or other attendants, particularly with inexperienced mares, which may become excessively possessive with a first foal. However, she will gradually allow other horses near her foal over the first few weeks and will, herself, start to socialise again.

She will experience uterine tension or slight pain as the uterus regains its tone and keeps contracting a little as it gets rid of any residual fluid leftovers from the birth. These processes may last two or three weeks, after which time the walls of the uterus will have regained much of their former tone and thickness (having become thinner as they stretched to accommodate the foetus and its associated fluids). The mare will regain her normal external appearance, her flanks and quarters will firm up again and she will gradually regain her figure.

The healing and regeneration process of the uterus and birth canal takes place fairly quickly in a healthy mare, as they must prepare for another oestrus cycle to start in readiness for another gestation for the following year's foal. If she is not mated again at her foal heat (a practice which seems to be becoming less popular), it allows more time for everything to return to normal, probably with a better chance of conception in a 'cleaner', stronger, rested uterus at the following heat.

The foal's milk requirement increases gradually over the early weeks until its demands are at their maximum by about two months after foaling. This places great nutritional demands on the mare, which must eat for herself and to produce most of her foal's nutrition for several months, until it can eat enough grass or other food to maintain itself. By this time, both mare and foal will probably be turned out or will be living permanently out with others, and the foal will have learned for certain who its dam is and what she looks, smells, tastes and sounds like. It will know that it can expect her to protect it from marauding mares (with or without their own foals), from other youngsters who might be over-enthusiastic or even bullying towards the newcomer, and to a lesser extent from the two-legged animals who seem to interfere so much.

Its eyesight will have improved greatly and it will be familiar with its paddock. It will be learning about the weather, night and day, warm sunshine, chilly breezes, rain, what its body and legs are capable of and all the other things that make life so fascinating and which fit it to be the prey animal it is.

Its body systems continue to develop and although at birth its legs were almost as long as the dam's (and by now it can run almost as fast as an adult horse) its body systems are starting to become really functional and it is starting to grow in earnest.

Its whole diet will be milk, but it may be seen eating adult horses' droppings, most likely those of its own dam, to equip itself with the intestinal micro-organisms which are essential to its future herbivorous diet. It will now be suckling for a few minutes at a time every two hours or so. Its baby coat will actually be starting to change to a sleeker, shiny adult one, and it might even be starting to change colour and look patchy and comical.

The foal is born with just the central incisor teeth in the upper and lower jaw (four teeth in all). By four weeks of age it will have its centrals, their neighbours the laterals (making a total of eight incisor teeth, four in each jaw) and the first three pairs of cheek or premolar (back) teeth, three teeth in each jaw, making a total of twelve 'grinding' teeth, all of which help in the foal's experiments with grass or the dam's hard feed.

'Outside' feed very gradually takes the place of the dam's milk. Although the foal may still be suckling her at the age of six months or older, if it is left with her, it will actually be taking little milk and less often. This naturally coincides with the time when, if she is pregnant again, her nutritional supplies will soon be going towards nurturing the new foetus within her during its final demanding few months in the womb – and then the process starts all over again.

Bonding and imprinting

Once a foal is on its feet and wandering around, particularly if the dam is a good broodmare and providing the right help and guidance, it will form an attachment to her very quickly. It is the act of suckling, when the foal learns the distinctive smell and taste of its dam's milk, that bonds it to her. However, so strong is the foal's instinct to survive that it will actually suckle any mare which lets it, even though it knows quite well she is not its dam. In practice, it is very rare for an equine dam to permit another's foal to suckle her and any little intruder will usually be pushed, bitten or kicked away without mercy, which sometimes results in injury. Because of this, a foal quickly learns that milk is only available from its own dam.

For her part, the dam should find her maternal instinct developing equally quickly. Pheromones in the amniotic fluid, which she will obviously smell before the foal is born, are believed to stimulate hormones which activate her emotions and instincts and 'stamp' their scent and therefore that of her foal, firmly in her senses and memory. Because of this, her affinity for her foal is partially in place before it is born and she will, under normal circumstances, mother only her own foal. (Persuading bereaved dams and orphaned foals to bond is dealt with in chapter 9.)

The traditional advice is to leave the mare and foal alone during foaling if all is going smoothly, or as soon as possible after an attended foaling so that they can bond with each other. They should be disturbed only for important tasks such as retrieving the afterbirth, checking the vital signs of health and wellbeing of both of them and, of course, providing the dam with feed and water. (This does not mean that a discreet eye should not be kept on the pair by means of closed circuit television or an observation hatch or window in the wall of the foaling box.)

The conventional view is that the presence of or handling by humans interferes with the bonding process between mare and foal. It is obviously vital that the foal realises that it is a horse and not a human, a common problem with hand-reared foals which are given no equine company. To ensure this, the very first thing the foal should sense and have much contact with after birth should be the dam and in the wild, of course, this would be so. It would be surrounded by its dam and, in due course, her herd-mates and family; other animals in the environment would obviously not be part of the family and may even be dangerous. The foal would quickly pick this up from the behaviour of the other horses, mainly the dam. In domestic situations, humans often cannot resist the temptation to attend even normal births which are going well and even to 'help' a mare which neither wants nor needs help. Although traditionalists will understandably frown on this, those who do it may unwittingly be part of a process called 'imprinting'. This is a form of learning which begins in a young animal's life once its eyes have opened. Species born with their eyes closed bond with their mothers meanwhile by means of smell, taste, touch and sound, but cannot imprint till their eyes have opened. As foals are born with their eyes open, bonding and imprinting occur at the same time. Imprinting is the mental attachment of young animals to (usually) moving objects within this sensitive period, during which they are particularly susceptible to absorbing environmental and social experiences which affect their learning behaviour probably for the rest of their lives. It was first described over sixty years ago by the Austrian scientist Konrad Lorenz, who imprinted himself on a clutch of newly hatched geese which then followed him everywhere as he took over the role of their mother, feeding them and even swimming with them – when they followed his head as it bobbed along.

The process has been recognised for generations and possibly for as long as humans have been training animals. It has been known that animals handled well from birth are much more accepting of humans and easier to train – but the scientific description of the process as a biological phenomenon belongs to this century.

In foals, the period during which imprinting can occur is from birth to one or at most two hours of age. After that time, they begin to show fear of moving objects, people or animals around them which, in the wild, could be predators. This is an excellent survival mechanism for an animal born in the open surrounded by its family and herd: even in the dim light of night when the foal would normally be born, it can discern its dam moving above it and will be bonding with and imprinting on her. By dawn, it may have to run with the herd away from danger and so, after the imprinting period, natural suspicions and fear of other animals take over.

The American veterinarian Robert M. Miller has devised a system he calls imprint training over a period of thirty years. He noted that the foals he delivered which needed veterinary treatment and a lot of handling at and immediately after birth were much easier to handle and train later on than those left alone with their dams. His system is described fully in his book *Imprint Training of the Newborn Foal* (see Appendix). Only an overview can be given here, but the system has been used for some years now, with such success in various types of horse that some explanation is called for.

The objectives of the system are to lessen or remove the trauma and stress of handling and training when the foal is young and as it grows. By coming to recognise humans as a normal part of their lives, foals are completely relaxed in their presence and Miller has found that the method also helps reassure inexperienced dams, particularly first-time dams which may be not only afraid of their foals but even aggressive towards them. If their trusted handlers are not afraid of them, why should they be? Miller is also very keen to remove the 'adversarial relationship' which so often exists between trainers and horses and replace it with one based on natural, automatic acceptance, submission, trust and willing co-operation and obedience.

The first aim is to handle the foal intensively and repeatedly in a ritualised way before it gets to its feet – immediately after birth. In this way, there is no need to actually 'intrude' on the mare as she gives birth (which puts off many mares) unless it is essential. The handler or trainer doing the imprinting (which must, incidentally be studied and done properly) lets the mare see what is happening and encourages her to bond with her foal, to lick, smell and touch it at the same time, and allows frequent short breaks for her to nuzzle and get used to it before starting again.

The foal is gently but irresistibly restrained on the ground while the handler repeatedly touches it all over (fifty times for every area is recommended), until it becomes completely desensitised to being touched in every area from nostrils to anus. The bottoms of the hooves are also patted fifty times to prepare for shoeing.

The whole procedure takes less than an hour, after which the foal is allowed to rise (around the time it would normally do so, anyway), find the udder and suckle. The restraint instills into the foal the knowledge that humans are dominant (which does not mean that they are a source of pain or fear).

In less than an hour, with Miller's techniques, the new foal can be desensitised to being tacked up, bitted and shod, having its ears, eyes, nostrils, mouth and tongue intensively handled, and having its feet, perineal or under-tail area touched. This is all in preparation for starting under saddle and for any veterinary, farriery and dentistry treatments.

By running the body of a buzzing electric clipper all over it, including the head and ears, it is desensitised to clipping, and can also be made 'immune' to bathing and sprays. The foal is obviously not actually tacked up, bathed or sprayed, but it is made familiar with actions and sounds that simulate those procedures.

Once the foal is up and steady on its feet, between twelve and thirty hours of age, the process is repeated. Further procedures are carried out to desensitise it to pressure in the girth area, flapping blankets, whirling ropes, a water hose and crackling plastic. It can also be familiarised with other animals and, by standing over it with no pressure on its back if the handler is tall enough, to actually being backed.

Foals can be desensitised to the things they will experience during their future careers, such as hunting horns, hounds if available, music, loudspeakers, gunfire, sirens, shouting and so on, and all this without frightening them.

In contrast to desensitisation, sensitisation is the creation of a conditioned response to a stimulus. The most famous example, which is nearly always used, is that of the Russian scientist, Pavlov, and the dogs he conditioned to salivate when they heard a bell. It occurs during the training process of horses, usually when they are some years old. Gradually, they respond without thinking to the trainer's various aids and signals.

Miller, however, sensitises day-old foals, aiming for five conditioned responses, each of which can be taught in minutes if it is properly done. They are:

- to pick up each foot when asked
- to be halter-broken to the extent that it will lead willingly and not pull back when tied (although it is not actually tied)
- to move the hindquarters sideways when asked
- to back in response to a vocal command and chest pressure
- to move forward in response to pressure on the buttocks.

It must be stressed that the foal is never frightened, forced, bullied or hurt. Because this training is carried out at such a young age, the foal accepts it without question, just as it does its dam's discipline and guidance. If it is done properly and thoroughly, with a few repetitions, the end result is a totally human-friendly, obedient horse, as respectful of humans as of a dominant horse. The horse will work willingly and enthusiastically as part of the human milieu and actually want to be with humans, accepting them as a normal part of its life, whilst still realising that it is undoubtedly a horse.

Further training procedures are added as the youngster matures and, for owners of aggressive or over-protective broodmares, Miller says this:

One of the most frequently expressed concerns regarding early foal training is that it will somehow interfere with the bonding between the mare and the foal, or that it will cause the mare to reject the foal . . . I have never seen a mare with an imprint-trained foal act aggressive towards humans.

He adds this caveat:

Mares that have had little human contact and are not, at least, tractable and well halter-broke are a different story. Their fear of humans will be transmitted to the foal. As I've said many times, no mare should be bred [mated] until she is halter-broke and well-mannered – and, ideally, broke to ride. Good horsemen aren't proud of unbroke or ill-mannered broodmares.

True pearls of undeniable wisdom.

Robert Miller predicted that 'more competent horse trainers than I' would adopt his method and I understand that the American trainer, Pat Parelli, is achieving 'astounding results' with the system. Miller says of Parelli's work: 'Seeing a skilled horseman use this method, with innovations and variations of his own, so effectively is very gratifying to me. It is, I believe, only the beginning of a new era in horsemanship.'

At a time when so much in horse management and riding is changing for the better, imprint training is a method which should surely receive the consideration of any breeder wishing to produce accepting, obedient, friendly, enthusiastic and respectful youngsters.

Chapter 7

Veterinary Care and Management of Mares, Foals and Youngstock

Lactation

The mare's udder has four compartments, quarters or glands, but the two glands on the left side of the udder exit through one teat, as do the two on the right, so the mare has only two teats. The udder, and especially the teats, have a good nerve supply, and are sensitive, and it is not uncommon for mares, and not necessarily only maiden mares, to dislike having their udders handled, washed etc., and for maiden mares to object strongly to their foal's efforts at suckling. Even after several foalings, some mares need to be tactfully restrained for the first few times their new foal suckles.

The udder has generous blood and lymph supplies. The quarters are separate, with no exchange of milk between them. The mammary tissue is composed of millions of tiny sacs or alveoli with linking ducts, arranged in lobes leading via ducts to a cavity or gland cistern in each quarter, from which run the teats. The milk is made in the alveoli in a lining of lactating cells. The nutrients needed for making milk are supplied in the blood in the form of small molecules which pass out of the capillaries (very fine blood vessels) into the lactating cells. The lactose, protein and lipid parts of the milk are synthesised in the cells and the milk passes out of the cells into the lumen or hollow centre of each alveolus.

The alveoli and ducts are surrounded by a network of myoepithelium, tissue composed of muscle cells which contract to 'bring down' the milk. The alveoli are 'serviced' by being surrounded by blood and lymph capillaries.

Young foals have very long legs and comparatively tiny, lightweight bodies which makes it easier for them to learn to manipulate their limbs without having too much bodyweight to control. (Vanessa Britton)

A number of hormones are needed to give optimal mammary development: prolactin (essential in all stages), oestrogens, progestogens, adrenal steroids, insulin, thyroid and growth hormones.

Milk production begins from about four weeks before foaling. After birth, as the foal's demands for milk increase, the dam's milk yield increases until, about two months after foaling, it is at its maximum; riding-horse dams will produce about 10–18 litres (2¼–4gal) of milk a day and ponies slightly more than half that much.

From around three weeks of age, most foals start imitating their dams and begin investigating other sources of food. The milk yield, and later its quality, both gradually decrease to encourage the foal in this. This leads naturally to weaning, in natural conditions or domestic circumstances where matters are allowed to more or less take their course, when the foal is about ten months old, although it is not unusual for older offspring, if they are still with their dams, still to take the occasional suckle, probably as a bonding or comfort action. Lactation naturally lasts almost a year, and the milk supply normally dries up a few weeks before the mare is due to foal again.

Commercial studs usually wean their foals at about six months of age, however, when the yield has already dropped considerably and the foal is taking nourishment from grass and its own supply of feed such as hay, haylage and concentrates supplied by man.

The mare's milk contains all the energy, fat, lactose (a sugar), vitamins, minerals and protein building blocks the foal needs during its early months. In addition, antibodies (immunoglobulin proteins) are also present, stimulated in the dam by any diseases with which she has come into contact, naturally or through vaccination. The foal's own immunity depends on absorbing these antibodies. It is normally advised that mares due to foal away from home are sent to stud four weeks before foaling so that their immune systems have a chance to develop antibodies to any diseases present in the area of the stud. Otherwise, she, and particularly the foal, which cannot develop its own antibodies for some time, will not have the right sort of protection.

Colostrum or first milk contains about 25 per cent dry matter, 16.5 per cent protein and 3 per cent lipid (fats/oils). Lactose concentrations rise during lactation, whereas total protein, minerals and energy levels fall slowly. Horse milk is lower in fat, energy and total protein and higher in lactose than that of most other species, so orphan foals should not normally be encouraged to suckle other lactating mammals such as cows or goats. Because little scientific work seems to have been done on this subject, it is not known whether there are any long-term physical effects on horses from being brought up on 'alien' milk.

As we have seen colostrum is essential, as the foal is born with no inherent immunity against disease and must obtain its protective antibodies from the dam via the colostrum. The foal's gut is only capable of absorbing these antibodies for up to twenty-four hours at the longest, so suckling early (ideally within six hours) is crucial.

The basic sequence of events which triggers lactation and sustains milk supply involves changes in levels of progesterone-related hormones, oestrogen, growth hormone and cortisol. This change allows the production of the hormone prolactin which is secreted by the anterior pituitary gland; it promotes the growth and development of mammary tissue, initiates milk production and maintains it after foaling, along with growth hormone and cortisol.

The hormone oxytocin (stored in and released from the posterior pituitary gland) effects the release or 'let-down' of milk by causing contraction of the smooth muscle in the alveoli of the udder, which pushes the milk down into the gland cistern and teat. Sometimes this causes mares to 'run milk' before and during foaling. (Oxytocin also causes contraction of the smooth muscle of the uterus during foaling.)

As the foal removes milk from the udder during suckling, it stimulates the production of more milk. Sensory (or afferent) nerves within the teats are stimulated and the mammary tissue contracts to bring down the milk. A nervous message also goes to the hypothalamus in the brain which then 'instructs' the posterior pituitary gland to secrete oxytocin into the bloodstream. It reaches the udder, where it also causes contraction of tissue which forces milk out of the alveoli via the ducts to the cisterns, from where the foal sucks it out of the udder.

Anything which upsets the mare can interfere with this process by causing the production of adrenaline, which brings about constriction of, among others, the blood vessels supplying the udder. This means that the levels of oxytocin delivered to the alveoli are reduced and there is consequently less of the tissue contraction which is necessary to force down the milk, resulting in a reduced milk supply. This is only one reason why broodmares must be kept calm, relaxed and content.

Periodic checks of mare and foal

Sometimes a veterinary surgeon will be involved in the health of a pregnant mare but may not be present at the birth (often because the foaling goes unobserved). The owner might also judge that since all is in order with the mare and newborn foal, there is no need to call in the vet in to check on them. However, at least for your first foal, it is probably wise to try to have the vet present for your own reassurance, in case something is wrong which you do not recognise. Read about the signs of imminent foaling (see chapters 6 and 17) and warn the veterinary practice as far in advance as possible. If the vet has been attending your expectant dam, you will both have a fair idea of when she is likely to foal.

Although breeding horses is not a job for novice horsemasters and anyone contemplating breeding should be knowledgeable about and experienced with horses in general, it is never a waste of time or money to call in the vet for a check at least just after foaling. The breeder should know enough to be able to spot any abnormality in mare and foal quickly so that the vet can be called early, when there is more chance of a disease or other condition being treated successfully. In addition to keeping an eye on the normal signs of health and disease, physical and behavioural, in the mare, the foal's temperature, pulse and respiration rates should be monitored (see chapter 6), and also its general demeanour. The normal behaviour is for foals to skip behind their dams at the approach of anything non-equine, as well as unfamiliar horses, and to suckle for reassurance and comfort. Foals are inquisitive and playful and show every sign of being energetic and interested in life. Diseases and abnormalities are dealt with in chapter 9, but basically, if there is *any* aspect of your mare's or foal's health or wellbeing about which you are doubtful, you should ring your vet at least for initial advice, and request a visit if necessary.

Mating at the foal heat

After foaling, your mare will probably come into oestrus again within about five to eight days. If your foal is born fairly late in the season or if you are involved in breeding Thoroughbreds or show animals which you want to look well developed,

you may want to have the mare served again at this time to get her in foal again as quickly as possible, leading to as early a foaling as possible the following year.

Increasingly, however, this practice is going out of favour, partly because conception rates at the foal heat are in general only about 50 per cent, and partly because many vets feel the uterus will not have recovered enough by this time to be able to accept and support another fertilised egg. This is especially likely if the mare had a difficult foaling. The uterus needs to return to its normal size and tone, and the mare may also have developed an infection, may have a retained afterbirth, may not begin cycling normally again or may not be producing ripe follicles.

If you feel that your mare must be mated at her foal heat, discuss the matter with the stud staff if she is away, and with the vet attending her. Hormonal treatment can be administered but generally, the feeling is increasingly that mares are best not mated at the foal heat. She should be in season again in three weeks, which is not long to wait unless, through repeated foalings every year, her foalings have become later and later each year, as they will. Perhaps a season off would be better for her and she may actually like to do some work.

Scouring (diarrhoea) in the foal at the dam's foal heat is not uncommon. It is believed to be caused by hormonal changes in the dam's milk at this time, which cause digestive upset in the foal. It is usually temporary but should not be treated lightly, as foals can soon become dehydrated by frequently passing loose or watery faeces. The veterinary surgeon should be called to check on the reason and to advise on treatment. It is not impossible, even at so young an age, for the foal to be suffering already from parasite infestation (*Strongyloides westeri* and/or *Parascaris equorum* – see page 64) or it could have picked up an infection. Whatever the cause, it should be established and suitable treatment given. Diarrhoea should not be regarded as normal at this time (it is not, although it is not unusual) and temporary. It is abnormal, a clear sign of something wrong, and needs veterinary advice.

Growth and development

Young horses, and foals especially, grow very quickly. As we have seen, a foal's legs are almost as long as its dam's at birth because, as a running, plains animal, it needs to be able to take

These older foals still show the long legs of their age but their bodies are noticeably bigger and longer than those of the very young foal. (Vanessa Britton)

long enough strides to keep up with the herd by the time dawn arrives, usually within a few hours of birth. Its body, neck and head appear very small in relation to the length of its legs and the size of its joints, although the hooves are small. The legs do, obviously, grow and mature, but most of the growth and development occur in the rest of the body.

Foals grow most quickly during the first month after birth. Newborn foals weigh about one-tenth of their adult weight and will have reached about half their adult weight by seven or eight months. During the first month, their height increases by about one-third. After this initial period, growth and development slow a little, but there is another surge of growth between six and twelve months, then another after puberty.

Older foals, in particular, have quite large hindquarters compared to the rest of their bodies, another natural provision for speed in the face of danger, as the hindquarters provide the propulsive power. Youngstock are normally croup high until about four years of age. Nature arranged for foals to be born during mild weather when food (milk and grass) is plentiful, and this enables these growth spurts (which are controlled by hormones) to take place; without the raw materials in the form of nutrients, the hormones' 'instructions' could not be fully carried out. In studies on Thoroughbred foals, it has been found that those born very early in the year are not as well developed at comparable ages as those born during a more natural time, and the difference can be still apparent at eighteen months of age and beyond – just the time when yearlings are being sold and put into training for the flat. (This is just one reason for abolishing the 1 January birthday, which was supposed to have the opposite effect; those who introduced it believed it would produce better-developed youngsters whereas the reverse appears to be the case. The other reason is economic; expenditure on labour, housing, feeding, heating, lighting, clothing and veterinary treatment are all necessary to bring animals into breeding condition at an unnaturally early time of year.)

There is no reason to suppose that the same interruption of horses' natural growth patterns does not happen in other breeds too, although they are less likely to be born in midwinter. It seems that, however scientifically correct the diet, it does not make up for good grass and quality mare's milk.

A horse cannot be said to be physically mature until the age of five, sometimes even six or later in late-maturing breeds or types such as Arabs and their crosses and big, rather backward individuals, and the vertebrae of the spine are one of the last parts to mature. The bones known as long bones (such as the pastern, cannon, forearm and thigh bones) grow in diameter in response to the maturing process and in response to stresses placed on them during work. There is a membrane called the periosteum (*peri* = around, *osteum* = bone) around the outside of the bone which produces new bone from bone-making cells called osteoblasts for this purpose.

Long bones grow in length from areas near their ends called epiphyseal plates or growth plates. The plates consist of cartilage (a softer, gristly type of bone); cartilage cells (chondrocytes) are produced in the area nearest to the ends of the bone (the knuckle part) and undergo several changes, finally becoming calcified and turning into new bone at the end of the plate nearest the shaft. One can think of the ends of the bone as being pushed away from the shaft as it lengthens due to new bone being formed in the plates. Eventually, the bone 'catches up' with the cartilage and the process (known as endochondral ossification) stops. At this point, the growth plate has become inactive and is said popularly to have 'closed'. This term was coined because some racehorse trainers have their two-year-olds' legs X-rayed to see if the plates above the knees towards the ends of the radius or forearm have stopped producing bone and have hardened, and, therefore, how much work the horses can probably withstand. On an X-ray, mature, 'solid' bone shows up as white whereas softer bone consisting more of cartilage is darker and makes the growth plates look open, hence the descriptions of growth plates as being 'open' or 'closed'.

The plates in the pastern and fetlock areas will have closed by six to nine months, those below the knees and hocks by about twelve months and those above the knees (with which flat-racehorse trainers are most concerned) by about twenty-four to thirty months. In the hindlegs, the growth plates in the femur or thigh (the largest bone in the horse's body which runs from the pelvis, at the hip joint, forwards and downwards to the stifle) and those in the tibia or second thigh/gaskin bones below the stifles will have closed at between twenty-four and thirty months; those above the hocks will close between about twelve and eighteen months.

Recent research seems to indicate that it may not be disadvantageous to work youngsters fairly hard, but judiciously, as bone and other tissues adapt according to the stresses placed upon them, although as in young human athletes, work to the point of physical (not to mention mental) distress – overwork – can result in permanent damage. If this subject particularly interests you, your veterinary surgeon will be the best person to keep you up to date with the latest research and thinking. Many warmblood and competition horse breeders do periodically work and jump yearlings and two-year-olds unridden, with a view to assessing their potential, so this subject is certainly also relevant to them. Potential event horses and steeplechasers are normally left longer to mature than other categories of competition horse so it is not so critical for them. But breeders of all types of horse and pony will probably find the subject of interest because the way a horse is managed and worked when it is young affects its abilities and constitution for the rest of its life – and even affects its lifespan, as most horses in domestic circumstances are put down if they can no longer work – except possibly for breeding stock, a subject discussed in chapter 13.

There are, of course, other factors involved in growth and development. Young foals of any breed are particularly susceptible to the adverse effects of cold, wet weather and also hot sunshine, and most concerned breeders take adequate steps to provide them with proper shelter. Even native ponies, cobs and heavy horses need *some* effective shelter and sensitive animals with any 'hot' blood in them at all should certainly have good shelter from extremes of weather at both ends of the scale. Those born at an unnatural time of year should have indoor (non-dusty) exercise facilities to give them the activity they need for the development not only of their bodies but also of their co-ordination and psychological welfare.

Disease can certainly arrest growth, as can poor nutrition (insufficient or imbalanced), worm infestation and distress. However, no amount of good management and feeding can make a horse grow beyond its genetic capacity.

From a genetic viewpoint, we do, of course, see humans and other animals which grow bigger with each succeeding generation. Our own species is an excellent example. If an animal has the genetic capacity to grow generally taller and bigger, good environment, feeding and freedom from disease and distress will encourage it. But if the capacity is not there in that individual, nothing we can do in the form of feeding and management will create it.

Castration

The choice of stock for breeding and deciding which animals should be castrated or gelded is covered in chapter 13. Castration is carried out to prevent a male horse passing on his genes (for whatever reason) or to make him easier to handle. Although it is usual for male racehorses, at least flat horses, to be left entire, and also competition dressage horses because of the extra presence the male hormones impart, most competition horses in other disciplines are castrated.

The operation which obviously involves the removal of both testicles, can be carried out at any age and some breeders habitually castrate male colts which are not regarded as good enough to breed from as soon as both testicles have descended. Most colts, however, are castrated between the ages of one and three years. Some breeders believe that leaving it as late as possible makes for better neck and crest development and a prouder temperament in the adult animal, but others feel that entire animals do not reach quite the same height as geldings and people often do not want the handling problems which can go with unbroken young entires.

For the animal's welfare, it is normally best to perform the operation in the spring, when the weather is mild and he can be turned out to exercise most of the time to assist in dispersing the resultant swelling and soreness. Do not wait until there are flies around, however, as they will make straight for the wound, and can irritate it greatly and cause infection.

The vet can perform the operation with the colt standing up, after giving him a local anaesthetic and sedative, or lying down under general anaesthesia. There are advantages and disadvantages in each method. With the colt standing, it is felt that there is greater safety for both horse and vet (there is always some risk with a general anaesthetic), but a general anaesthetic allows more time and it is easier for the vet to see what he or she is doing.

After the operation, the colt may lose some condition, but this should soon be put back with correct feeding. The male hormones do not subside immediately; it may take a few weeks for the colt to stop thinking and behaving like an entire, during which time he will still be capable of covering and impregnating a mare, owing to the presence of a residue of viable sperm in the spermatic cord.

An old practice which has not entirely died out was for breeders to ask for the colt to be 'cut proud'. This involves leaving a little testicular tissue to help the horse retain some extra 'spark' and presence. Such geldings often retain stallion-like characteristics and can be a real nuisance. Most vets would refuse to do this as it really constitutes faulty surgical technique and, of course, can also make life unpleasant for the horse.

Vaccination

Your mare's vaccinations should always be kept up to date, and she should certainly have completed a full vaccination programme against any diseases your vet has advised according to prevailing conditions in your region or country before she goes away to stud. Some diseases currently vaccinated against, according to country and conditions, include equine influenza, tetanus, strangles, rabies, equine viral arteritis, *Brucella abortus*, African horse sickness, equine encephalomyelitis and equine herpes virus. Your mare should have completed her course, or been given her boosters, a month before the date on which she is due to foal. The vet is in the best position to advise on current strains of diseases, appropriate vaccines and a suitable programme of initial doses and boosters. The immunity she acquires will be passed on to the foal in her colostrum, after which it can start its own course at four months of age.

Most studs will not accept unvaccinated mares and will ask to see their vaccination certificates. Your mare will also probably be mixing with other mares from different regions so it is best all round for vaccinations to be up to date.

Swab tests

Responsible studs will also insist that your mare is swabbed to confirm whether or not she has any infectious disease of the reproductive tract. If she is infected, not only will the sperm and eggs not survive for long, the mare is also unlikely to become pregnant or maintain a

Foals need a lot of rest and sleep so will spend much time lying down. An experienced bovine mother regards one of her equine fieldmates with interest

pregnancy, and she will pass the disease on to the stallion, which will infect his other mares. This can have disastrous consequences, financial and otherwise, for all involved.

The vet will take a sample of material from the mare's clitoral fossa, vagina and/or cervix via a speculum onto a swab. This is cultured (grown) in a laboratory and the results should be available in about a week. It is advisable to take these tests (on your vet's advice) well before the mare is due to be covered so that there is time for any treatment to be administered and as few heats as possible are missed.

Signs of disease (the absence of which does not mean that the mare is not infected) can be observed at any time, but are most easily seen when the mare is in season, particularly in the case of uterine infections, as the cervix is then open and any discharge can reach her vulva. Any excessive discharge from the vagina, around the vulva and matting the tail hairs together – sometimes even caked on the insides of the hind legs – indicates a likelihood of infection. The vet will need to wash out the mare and give her antibiotics to treat any bacterial infection. Fungal infections are less common.

An ultrasound scan can detect whether or not the uterus contains accumulated fluid from infection (in older mares, particularly, the uterus can sink somewhat, making fluid retention more likely). The administration of the hormone oxytocin, which stimulates contraction of smooth muscle such as that in the uterine wall, will cause the uterus to expel excess fluid and so help in the treatment of infection.

Worming

Infestation with internal parasites is a major cause of unthriftiness and illness in horses and ponies, and the damage parasites cause can result in their deaths. In practice, it seems to be impossible to maintain animals completely free from parasites. A common way of gauging how much of an infestation a horse may be carrying is to have a faecal count done in a laboratory (something which is usually arranged through one's veterinary surgeon). The numbers of eggs of relevant species are counted under a microscope and if an animal has less than fifty eggs per gram of manure, it is regarded as being as free as is practically possible. Another way of checking is through a blood test, which will show up the presence of alien proteins. The orthodox advice on worming horses is to work out with one's vet a proper worming protocol, follow it faithfully and take regular checks (through droppings or blood tests) to keep tabs on the level of any infestation. It is important to determine the types and levels of parasites present so that you can be sure you are using the right drugs for the worms your horse has, at the right times of year and at the correct dosing intervals, otherwise you could be endangering the health of your animals and wasting your money.

Another important step is to keep the environment, particularly the pastures, as free from manure as possible. During warm, moist weather, it is advisable for droppings to be removed from paddocks at least every three days and preferably more often. Many well-run establishments remove them daily, as a build-up of several days' droppings from a few horses can be a Herculean task to remove manually. Vacuum machines designed for the job can make life much easier.

This is the orthodox advice but in an age when so-called natural medicine and management practices are becoming more and more popular, other opinions should be mentioned. Some vets and breeders are opposed in principle to the regular dosing of animals, particularly breeding stock, with synthetic drugs which, by their very nature, are toxic although tested as safe when used according to instructions. They believe one should only dose individual animals if their tests show they need it. Veterinary surgeons who are also qualified in homoeopathy and other complementary therapies may prescribe other remedies to help deal with parasite infestation. As for picking up droppings from paddocks, the reason for this is that droppings contain worm eggs and larvae which contaminate the grass and are eaten by the horses, thus reinfesting them. If the droppings are removed as scrupulously as possible before the larvae hatch out and become infective, the horses will remain as clear as possible of parasites; picking up droppings has been described as the most effective practical step we can take to keep horses free from worms.

Studies have shown that horses correctly wormed but turned onto paddocks from which the droppings were not picked up regularly showed high faecal worm counts whereas those that had never been wormed but turned on to paddocks from which the droppings were removed daily were relatively clear. However, if you wish to follow this practice, very regular faecal counts should be done on your horses.

There are also those who believe that natural land management requires that droppings be allowed to rest where they drop, to rot down and fertilise the land. In natural conditions, and on establishments where mixed grazing is practised, where the animals have large areas over which to roam and which have many paddocks through which they can rotate their stock, this may well be a safe policy to follow, but most owners and breeders of small numbers of horses or ponies are not so fortunate. Land which carries significant numbers of horses soon becomes

Although horses are herd animals and need each other's company, they need to learn fairly early in life to regard humans as extensions of their equine family and to go with and trust them. Here, a yearling on a Thoroughbred stud gets to know his environment with his handler.

overgrazed and overloaded with droppings if they are not picked up more or less daily, and this must be a health and hygiene hazard to the animals grazing it. (It is conventionally said that you need at least 1 hectare (2 acres) for one horse with half as much again for each additional horse, slightly more for a mare-and-foal pair, less for ponies.) As well as worms there is a problem with flies in summer, which are attracted by droppings and breed furiously, laying their eggs in the piles (and muck heaps) and causing misery and possible ill-health to the horses.

In any case, this practice (which can never be truly natural in domestic circumstances) ideally demands mixed-stock grazing, with cattle as the most favoured complementary species for grazing before or with horses, and sheep the next most favoured, grazing with or after the horses. Cattle cannot graze short grass; they graze longer herbage and the horses follow, eating the shorter material. Horses are quite capable of grazing off grass right down to the soil, as are sheep.

From the worm-control point of view, it is a natural control, which is only partially effective, for a species normally to dislike grazing near their own droppings. In this way, it is hoped that they will miss picking up most of the larvae which have hatched from the droppings of their own kind. Cattle, sheep and horses happily graze over each others' droppings, however, ingesting the parasites of the other species, which cannot live in an unnatural host. In this way large numbers are killed off.

In modern farming and horse-breeding, however, particularly where cattle are concerned, one has to be careful about multi-species grazing because of the products (drugs, hormones and antibiotics etc.) which are routinely fed to cattle and which are passed out in their faeces to the land. Sheep are dipped with organophosphate products to protect against skin parasites and these can also sometimes cause environmental and health problems.

These are all matters to be discussed carefully with your vet before you reach a decision about which route to take. The facts are that significant parasite infestations seriously affect the health, wellbeing and even the lives of your horses and ponies, and probably the best any of us can ultimately do is to follow sound professional advice, as well as our own inclinations, to ensure that our horses and ponies are kept as free from them as possible.

As a matter of interest, some fascinating and highly relevant research was carried out in the 1970s at the then Equine Research Station (the Animal Health Trust) in Newmarket, England, by Dr Marytavy Archer on grazing preferences, use and grassland management for horses. It seems that this work was never continued, which is a great pity as it was showing some most interesting results.

Dr Archer concluded that the ideal horse to cattle stocking ratio for horse paddocks was one horse and ten cows to the acre (½ hectare) on good land to maintain the land in good heart and counteract the horses' tendency to graze land unevenly, producing short 'lawns' on the horses' favoured grazing areas with longer, rank 'roughs' near their own designated lavatory areas, which contained their droppings.

She also found that horses tend to be somewhat lazy and go to the lavatory area to dung but do so on the edge of it, rather than pick their way through distasteful droppings, long grass and thriving weeds: gradually, this area thus becomes bigger and bigger while the favoured grazing areas grow smaller and smaller. Weeds also tended to proliferate on land left to itself with the result that even after many months' rest, the horses, when returned to that paddock, kept to their previous lawn and lavatory areas, and so the process continued.

Very significantly, as far as palatability was concerned (and even the best grass is no good if the animals will not eat it), Dr Archer found that the horse droppings contaminated the land with their smell within half an hour of a fresh pile being deposited and the smell remained, putting off the horses, even if the droppings were removed daily. To prevent contamination, it seemed that the droppings had to be picked up within half an hour of their being voided, something quite impossible for most studs.

She found, however, that spreading cattle manure (she used old-type farmyard manure based on straw) disguised the smell even if the droppings had been lying for weeks or months, so that, when they were returned to the paddock after its rest, the horses often formed different grazing and lavatory areas. This helped the grass species in the paddock to tiller (spread) evenly and helped keep down the weeds, which found it hard to compete on tight, close-bottomed land against healthy, growing and spreading grasses.

Although opinions, fashions and practices change over the years, we must all know from our own observations that leaving horse droppings down does have the effects reported by Dr Archer, even if we have not been sure why.

63

As I have said, if you have a great deal of land and can rotate, graze with other species and give your land long rests, you probably do not need to engage in the very time-consuming and back-breaking chore of picking up droppings. Even if you use a motorised buggy with a suction device made for doing this task, it cannot be denied that the carbon monoxide exhaust emissions will regularly contaminate your land (as they will from any motorised vehicle, including a tractor) unless the vehicle is battery operated, and I know of no such machine currently available. Spreading droppings, Dr Archer told me, did help desiccate them on hot, dry days but they must only be spread within the lavatory areas otherwise they will infest the lawns. It is much safer to actually remove them than to spread them into the wrong areas.

So testing and treating the horses, picking up droppings, rotating and resting your land and, if possible, cross-species grazing, can all significantly help keep down the levels of parasites in your horses. Spreading cattle manure in the form of slurry would presumably have the same effect of disguising the smell as using straw-based farmyard manure, but bear in mind the comments about drugs etc. above. Dr Archer stressed, however, that if pig or poultry manure is used on horse paddocks a hay crop must be taken before allowing the horses to graze, as otherwise the resultant high nitrogen content of the grass will cause metabolic and developmental problems in the horses.

Because anthelmintics (worming medicines) are continually being developed and tested, and it is hoped that, in due course, an actual vaccine may become available, specific drugs are not recommended in this book. Your veterinary surgeon is the best person to keep you updated on what are currently the most effective drugs and treatment regimes.

The first internal parasite a very young foal is likely to encounter and which can be passed to it from the dam's milk is *Strongyloides westeri* or threadworm. Although it is one of the causes of diarrhoea in young foals, as well as of poor appetite and condition and of lethargy, it is not regarded as being of great significance. The adult worms live in the mare's udder and the foal ingests larvae when suckling, so it can become infected during its first days of life. The larvae quickly mature, so that eggs can be detected in the foal's droppings from about two weeks of age; between six and nine months, it appears to develop an immunity to *Strongyloides westeri*, when the infestation disappears. However, if unchecked, this parasite can become very numerous and cause chronic rather than transient diarrhoea and very poor condition and health in the foal. It is easily treated with conventional drugs.

A parasite which can cause much more trouble in foals than the threadworm is the roundworm, *Parascaris equorum* or *Ascaris equi/equorum*, otherwise known as the ascarid or large whiteworm. It is most dangerous in foals up to six months of age but continues to cause trouble up to about eighteen months, by which time good immunity should have built up, making the gut environment less favourable for the worms. Until then, however, large numbers of eggs can be produced and ingested, leading to heavily infested pasture and youngsters.

The animals take in the eggs with grass or from licking objects (including stables and fencing) with the eggs on them, even their dams' and other horses' hind legs. The eggs have three resilient coats which protect the larvae, sometimes for years at temperatures below 10°C (20°F). In warmer weather, the larvae hatch inside the eggs within ten days and, once inside the foal or horse, hatch out and pass through the gut wall to the liver, travelling on to the heart and into the bloodstream and lungs. From the air passages, they travel up to the throat, are swallowed and pass back to the intestines, developing into adults in the large intestines. Here they lay eggs which are passed out in the droppings, and the approximately eight-week cycle begins again.

The larvae damage the tissues of the liver and lungs during their migrations, causing poor condition and possibly pneumonia. As they can reach the size of earthworms when mature and in a good environment, roundworms may cause blockage or even rupture of the small intestine. Easily recognised symptoms of the effects of these worms include coughing, fever, weight-loss, poor condition and general ill health and failure to thrive.

Both threadworms and roundworms should be eliminated in youngsters, starting from four weeks of age (or before if your vet advises it). Broodmares should be wormed (for any species of parasite) before foaling and their udders, hind legs and tails kept washed and clean to help remove the eggs with their sticky, outer coat. Because the eggs can stick to surfaces and equipment, special attention should be paid to hygiene, power-cleaning the stable, disinfecting premises and keeping the bedding fully mucked out and clean.

Once the foal starts to graze more, the main parasitic danger will be from redworms, so called because the blood they suck from the gut wall gives them this colour. There are several types of redworm, the most significant for horse owners and breeders being the large strongyles or redworms (*Strongylus vulgaris* and, to a lesser extent, *Strongylus edentatus*) and the highly dangerous small redworms (*Cyathostomes*).

The large redworms are ingested from pasture as larvae and pass into the small intestine. *S. vulgaris* larvae penetrate artery walls and travel to the aorta, the main artery of the body. From here, they develop in the arteries supplying the small intestine and, by causing blockages or even aneurisms (ballooning and possible rupture of the arteries), they disturb or even cut off the blood supply to the gut, sometimes causing the death of part of the gut. This, in turn, causes potentially fatal colic as digestion is seriously disturbed. From the arteries, the larvae return to the colon, mature and begin laying eggs. The life cycle can take from two to twelve months.

The larvae of *S. edentatus* enter the peritoneum (the membrane lining the abdomen) rather than the bloodstream. Here, they form nodules which may bleed and damage the peritoneum. Small redworms (*cyathostomes*) are also taken in from grass as larvae and cause a great deal of gut damage by burrowing into the mucous membrane lining the large intestine (caecum/colon) forming cysts in which they develop further during the autumn. They damage the glands in the gut wall and also reduce its motility, so significantly interfering with digestion and causing colic. In spring, vast numbers of them emerge from their cysts all about the same time, causing extensive, potentially fatal, damage, inflammation and bleeding. The horse may have a fever or diarrhoea, and will lose weight rapidly. It will be depressed, dehydrated and have a poor appetite and may well die if veterinary treatment is not given at once.

With all these types of parasites, treatment with conventional drugs is effective and is currently aimed at killing the dangerous larval forms as well as the mature worms; work continues on producing effective drugs. Because of the dangers of infestation with the parasites mentioned in particular, and the great damage they can do to the internal organs and blood vessels, it is essential that veterinary advice is sought on the prevention, as far as possible, of parasite infestation, good land management, correct treatment and regular monitoring of the efficacy of the regime. There are certain drugs which must not be given to pregnant or lactating broodmares and, as usual, your veterinary surgeon can advise on safe, effective medicines for all your breeding stock.

Other parasites do not cause as much trouble but must still be eradicated. Pin or seat worms (*Oxyuris equi*) live in the large intestine, having been taken in with grass, and, after mating, the female worm crawls through the mare's anus, laying eggs on the perineal area (under the tail, around the anus and vulva). The eggs drop to the ground about three days after being laid and develop on the grass into infective larvae. The main problem pinworms cause is irritation of the rectum and perineum, sometimes severe.

Tape worms (*Anoplocephala magna* and *A. pertoliata*) are not very common in horses but must still be watched for. The type which infests horses is not the very long type which can be found in humans but is still segmented, flat and can be up to 80cm (30in) long. The segments, which contain eggs, are passed out in the dung and the eggs are eaten by forage mites found on the pasture and also on hay and straw. They develop inside the mites into an intermediate larval stage called cysticercoids, are eaten by the horse and, after about six to ten weeks, develop into breeding adults. They can cause blockages of the gut and, therefore, colic.

Bots are not worms but are still parasites. They are the larvae of the botfly (*Gasterophilus intestinalis*), which lays its eggs on the horse's neck, shoulders and forelegs in early summer and autumn. When the horse, or another, licks the area, the eggs are taken into the mouth and the larvae pass through the lips or tongue to the stomach where they overwinter as plump maggots, causing discomfort or pain, ulceration and digestive disturbance, before passing out in the droppings in the spring. The larvae pupate and the adult flies appear and start breeding, to begin the cycle again.

Treating horses for bots after the first frost of the autumn (which will have killed off any remaining adults and any larvae which may have passed through onto the land without attaching to the stomach lining) and again in spring breaks the cycle. If all owners did this without fail the bot fly would become extinct in a year, in a similar way to the way the warble fly, which occasionally attacked horses, was exterminated in cattle. The use of effective insecticides and repellants which kill flies and prevent them landing on the horse, prevents the eggs being laid in the first place and this, again, would be a virtually foolproof way of eliminating bots if all owners did it at the same time.

Farriery

Many owners overlook the need for good foot care for their mares. It is not uncommon to see overweight and pregnant mares tottering around on long, overgrown, cracked feet and to be told that their feet are attended to without fail every quarter, as if this were something to be proud of. But although broodmares do not perform fast, athletic exercise in the way of performance or competition work, their legs and feet are still under strain because of the greater weight they carry for much of the year. This weight is also carried twenty-four hours a day, whereas a performance horse's periods of stress and strain usually comprise only an hour or two per day, if that. Badly balanced feet which have been allowed to become lopsided and/or low in the heels and long in the toe, as well as cracked and chipped, can cause the mare pain in her feet and legs as excessive and uneven weight is borne on the bones and soft tissues of both. Laminitis due to overlong feet is common in broodmares. Pain makes the mare reluctant to take the exercise which is essential to the health of herself and her developing foetus, and may even cause her to lie down too much. In addition, it can stimulate the production of stress hormones which can adversely change the body chemistry and metabolism, which is again bad for the health of both mare and foetus.

Whether your mare is shod or not (and this will depend not only on her nutrition but also on her individual constitution, which will dictate her horn quality), she still needs to have her feet kept properly balanced as if she were working. The matter of shoes depends on the quality of horn she produces and the ground she is on. It is probably good practice to keep pregnant mares shod in front (as the front legs carry about two thirds of the weight) but not behind, but much depends on individual circumstances. Good farriery practice dictates that she should have her feet attended to probably every five to seven weeks depending on her rate of horn growth. If she is shod, her shoes should be removed before foaling to prevent possible injury to the foal.

Foals are often born with excess soft lobes of horn growing from under their hooves, called golden hoof. This will have accumulated because of the lack of friction in the uterus to wear it away. There is no need for it to be trimmed off; it is very soft and will even wear away on the bedding or as the foal is led with its dam from stable to paddock, on the ground and the turf. Foals' feet are not wider at the base than the coronet like the feet of mature horses, and novice breeders may be concerned about this. As the hoof grows and bears weight, a more familiar shape will appear. The farrier will probably aim to trim the feet in such a way that the frogs underneath are central and the feet, joints and legs are well in line so that the foal stands up with a correct hoof/pastern axis.

Corrective trimming to put right 'crooked' feet and legs can be undertaken quite successfully until the foal is six months of age whilst the bones of the feet and legs are still fairly soft and malleable. After this age, the process is much more difficult. For this reason, although some irregularities may right themselves, it is always worth getting your vet or farrier to look the foal over to check his foot and leg conformation. They may well spot something you have not recognised. Even a slight twist or unevenness in a leg or foot can become a real problem if nothing is done early enough to correct it.

Foals normally need their first trim at around two to three months, but before this you should prepare them for their first visit by very gradually and gently, but insistently, handling the feet and legs, even for just a few seconds whenever the foal receives any other attention. Although farriers and vets are trained to handle difficult animals, teaching horses manners is not their job and life will be easier for everyone, including the foal, if the breeder makes the effort to produce well-mannered foals.

Dental care

At present, there is no qualification in equine dentistry in the UK although the matter is under review. There are people who are not vets, but who do an excellent job of rasping horses' teeth, but even if you have one of these to attend normally, it is as well to remember that should it be necessary to remove a tooth or perform any procedure which is actually surgical the law in the UK requires that it must be done by a veterinary surgeon. Also, only a vet, or the owner under veterinary instructions, may administer a sedative to a horse which may need calming to have its teeth rasped.

In the USA, the World Wide Association of Equine Dentistry offers an excellent course and qualification in equine dentistry but their qualification does not entitle holders to practise in the UK except on referral through a veterinary surgeon.

Broodmares should certainly have their teeth checked once a year, maybe twice if they are older. They need to be able to eat without discomfort and to make the best of their food, so their teeth must be in just as good condition as those of working horses. The tables of the cheek (grinding) teeth need to be level, the sharp edges which form on the outsides of the upper teeth and the insides of the lower ones need rasping off, and any hooks which may form on the fronts of the upper cheek teeth and the backs of the lower teeth must be removed. Obviously, if any teeth are diseased or broken and need to be removed, a constant watch must be kept on the opposing tooth which will grow into the gap left and prevent a proper chewing movement.

As far as the incisors are concerned, mares which are parrot-mouthed or, conversely, undershot, may not be able to grip the grass and graze properly so time should be spent watching them eat to see if they are coping. In fact, mares with these defects should not be bred from as they can be passed on.

Any problem which prevents efficient mastication of food can lead to indigestion and colic, and the last thing you want is a pregnant or suckling mare with colic, as this can endanger both her and her foal's lives.

As part of the foal's veterinary check after birth, the vet will probably also wish to check its teeth. At birth, it will have four central incisor teeth (two each, top and bottom) and possibly also six pairs of milk cheek teeth, which will otherwise appear by the time it is a month old. The vet may wish to check that the foal's incisors are going to meet properly, that it has no deformity of the bone structure and that it is neither parrot-mouthed nor undershot.

Horses do not acquire a full mouth of permanent teeth until the age of five years. During the teething process, the milk teeth may become jammed on top of the permanent teeth erupting in their place. These remnants of the milk teeth are then known as caps. They can cause a good deal of discomfort to the horse and prevent it eating properly, so your vet will have to remove them, normally a simple procedure with special pliers.

Youngsters can get dental caries (decay) and also sustain broken teeth, either from kicks or from biting on something hard such as a stone picked up with grass or in hay or haylage. Painful lumps often appear under the lower jaw during the teething process as teeth develop and erupt, and a similar process probably occurs in the upper jaw where we cannot see the lumps. As the horse will be bitted at some time during the teething process, it should be remembered that the mouth may be painful and allowances made for this. Tight nosebands and restrainer headcollars and halters can also exacerbate pain, and getting tough with the youngster is not the way to handle this situation. Time should be allowed for the natural process to occur. Youngstock should have their mouths examined preferably every three months to check that all is in order and proceeding normally. If not, appropriate treatment can be given.

Back to work?

Some owners of only one or two broodmares, especially those who are not intending to breed more than one or two foals, often want to know if it will do any harm to ride a mare when she is suckling a foal, and opinions on this vary.

Once the foal is a few months old, say four or five months old, many people feel it does no harm for the mare and foal to start being separated for just a few minutes, perhaps leaving the foal in the box while the mare is groomed outside and possibly walked round the yard. This may be extended to separations of up to half an hour and it may be a good time for expert handlers to take the foal for its own little walk around the premises if it has been handled properly and is leading well.

If it is decided to exercise the mare, she should be restricted to very gentle exercise, as otherwise her milk supply will be adversely affected. Some people do ride their mares around and let the foal follow, but I have known terrible accidents occur this way. And if the foal is left in a box while the dam is out on a short foray (it should never be left in a paddock), it would be best for someone to remain with it, talking to it, handling it and trying to get it interested in some hay or other food. Remember that undue stress to either mare or foal will do neither of them any good and the whole process may well not be worth the risk, depending on the temperaments of the pair involved and the skill of the handlers.

67

Chapter 8

Nutritional Requirements of Mares, Foals and Youngstock

The correct nutrition of breeding stock is crucial to health, production and development. In fact, scientific knowledge of nutrition and the practical art of feeding together comprise the single most important subject in the whole field of horse management.

Getting the right nutrients, *in the right form*, into a horse makes an unbelievable difference to its condition and abilities. The phrase 'in the right form' is important because it is quite possible to formulate a diet which, from the nutritional, scientific point of view, contains all the nutrients a horse may need so that, in theory, it ought to be able to live perfectly well on it, but which in practice does not answer the horse's needs.

Mature horses' digestive systems are adapted to process fibre; they are not good at processing cereal-grain concentrates to any great extent or at working effectively on any diet which does not take sufficient account of the need for bulk. This bulk is essential for the system to work properly and for the horse to feel physically comfortable, well and psychologically content. The attempts some years ago of feed manufacturers to provide a feed, usually in the form of so-called 'complete horse cubes', which did away with the need for hay, were relatively short-lived. Not only further research but also concerned owners' practical experience proved that horses on such diets were miserable, bored, stressed and uncomfortable, and exhibited all the other signs one would today expect of a horse denied adequate bulky roughage. The cubes were said to expand once inside the horse, giving it that essential feeling of fullness, but that plainly did not happen. So here

For the first few months of its life, by far the most important source of nutrients for a foal is its dam's milk. (Vanessa Britton)

was a feed which contained the right nutrients for horses but in quite the wrong form and which was, therefore, in many cases actually damaging to horses' wellbeing.

Of course, horses' teeth have to be in good order and horses have to be as free as possible from parasite infestation in order to make the best use of their food. The food also has to be palatable to the horse as otherwise, no matter how ideal it may appear and no matter how delicious and irresistible the owner thinks it must be, it is useless if the horse will not eat it. Most horses, even those which are regarded as greedy, will not eat food they do not like, so the art of pleasing them is just as important as the scientific expertise which has gone into creating the feed.

Given these three primary conditions and assuming there is nothing clinically wrong with the horse to prevent it metabolising feed properly, you will find that feeding alone constitutes over half your job of maintaining your animals in healthy condition. With a good diet, the raw materials for growth, development, maintenance and energy are met, the immune system has every chance of working optimally, and horses feel good, look good even in the rough, behave better and have an improved attitude to life (temperament permitting); they perform well at work and at stud and are generally a lot less trouble than badly maintained horses which are either under- or over fed, with all the problems those two conditions can bring.

The science of nutrition, which is still incomplete but improving, is contantly coming up with new information on what nutrients the horse needs and how it uses what it eats; research goes on apace and the more reputable feed and supplement companies formulate their products accordingly and give their consumers accurate, unbiased advice. But science does not have all the answers, and the art of feeding horses, preparing their feeds, presenting them appropriately and knowing by experience, common sense – and also that vital element intuition – just what a particular horse seems to need, is essential to health, growth, development and wellbeing.

The right advice

In the past decade or so, a new type of specialist has become familiar to most conscientious horse owners – the equine nutritionist. The economic climate in the UK has made it very difficult for consultants of this type to make an independent living. Most nutritionists are employed by feed companies to formulate their products and, depending on the company's activities, to advise consumers directly, usually on telephone helplines where you can speak to an equine nutritionist free of charge. They will obviously advise on their firm's products but are also now very willing to give general advice on feeding and nutrition.

Indeed, advice is becoming less biased towards a particular company's products: it is not uncommon for callers to be given information on other firms' feeds or supplements, where appropriate; this may be just a public relations strategy, but one hopes it is a genuine concern for equine wellbeing. Formerly, horse owners found it necessary to ring several firms for advice, each of which promoted its own products for obvious commercial reasons, and they were often more confused after several hours on the telephone than they had been before they started!

There is no actual qualification in equine nutrition as such but there are now several universities, and colleges associated with them, which offer higher educational qualifications, including first and higher degrees, in equine studies and equine sciences, and which deal with nutrition in great depth and are regarded as suitable to qualify their holders to work as equine nutritionists.

Veterinary surgeons are also obviously well qualified to offer advice on nutrition, as the subject is amply covered on their qualifying courses. Some vets are as willing to work with nutritionists as they are with farriers, physiotherapists and other specialists. Whereas vets may not have specific knowledge of individual feeds or supplements on the general feed market, they can certainly look at the formulation of any product and tell you whether or not it is suitable for your horse's needs. Should your horse have any clinical disorder stemming from poor nutrition (usually from imbalances or specific deficiencies or overdoses), your vet is certainly the person who will have to put it right and advise you on how to feed the horse in future. He or she will probably be able to provide you with appropriate products (not feeds) and any supplements needed, some of which may only be available through a veterinary practice.

So good advice is readily available and commercially available, branded feeds made by reputable companies are better than ever before, including bagged or baled forage feeds as well as concentrates. Good feeds have the analysis on the bag or bale; if they do not, ask the manufacturers for full details. Most good feed companies have accurate, understandable literature on their products, which should help you choose the right feed for your horses, perhaps along with advice from their nutritionist. Any feeding recommendations as regards quantity are a guide only so here the art of common sense comes into play, plus the vital ability to know your horse.

Supplements

The subject of supplements is a positive minefield. In the UK, unlike the USA, there is currently no requirement for a product to be scientifically tested as even safe, let alone effective or suitable for its advertised purpose, although any product has to be suitable for the job for which it was bought and there is a certain amount of protection for consumers under the law against faulty or dangerous products.

I believe it important to take expert advice from a veterinary surgeon or a well-qualified, unbiased nutritionist before feeding any nutritional supplement, for whatever purpose. Most horses fed a varied, well-balanced diet do not need vitamin and mineral supplements. It should also be remembered that overdoses can be much more dangerous than deficiencies. Some products do have the backing and evidence of scientific trials and it is a good idea to ask for such data to be sent to you so that you can discuss it with your vet.

Herbs, being plants, are food to a horse, and in an ideal world, horses would be able to roam and pick and choose, using their nutritional intuition to choose what they want and can find. For many years owners have been recommended to sow a herb strip in horse paddocks, but owners who keep their horses on other people's property (which means most of them) are not usually in a position to do this. Failing this, many owners feed a herbal supplement to make up for the lack. The amount of documented evidence for herbal supplements and their effects is growing. The British Herbal Medicine Association has produced the scientifically based British Herbal Pharmacopoeia and has a very active Scientific Committee. In addition, qualifications up to degree level are available in phytotherapy (herbalism) and are increasingly being taken up

by those wishing to qualify in this field.

The art and science of herbalism is the oldest form of therapy in the world for both humans and animals and many substances derived from herbs and other plants are used in modern, orthodox human and veterinary medicine. Horses and their ancestors have survived for many millions of years and about three-quarters of that time has been spent eating. It is perfectly reasonable to suppose that they have been medicating themselves with herbs and other plants, including grasses, since long before we came on the scene and started controlling their diets.

The law in the UK on nutritional remedies leaves much to be desired at present, and herbal remedies must be sold only as nutritional supplements, without any clear statement as to what they are intended to treat. However, the unavoidably nebulous wording on the pack will give potential purchasers some idea, and the sensible thing is to follow the instructions or consult a herbalist as to how to use them. Moreover, the ingredients must be given on the pack, so it is a simple matter to look up the plant ingredient in any good book on herbalism to see what conditions that plant has traditionally been used to treat. Where many people may go wrong is in trying to be their own vet and trying to treat conditions they believe the horse has, or could have, by buying and using a supplement (herbal or otherwise). If the horse has any disorder, the first thing to do, unless it is something very minor with which you are sure you can certainly deal, is to contact the vet for a consultation and diagnosis. Once the vet (the only person, under UK law, able to diagnose an ailment in an animal) has stated what appears to be wrong with the animal, you can back up the treatment with appropriate herbal therapy. Most vets would not object to this even if they thought you were wasting your money! Increasingly, some would be quite happy about it and a few actually in favour of it.

As we in the UK already live in a nanny state in which 'Nanny' does not always know best, I am loath to recommend even more restrictions on our personal freedom, but it is true that many people believe the whole situation about feeding and therapy is in need of a thorough shake-up, particularly because our animals are usually no longer able to choose for themselves what they eat but end up eating what we give them – if they will. The British Equestrian Trade Association has formed a feed sub-committee to

Ponies and cobs evolved to thrive on 'poor' grasses. Unless the keep is excessively poor, they will probably only need extra feed during winter.
(Vanessa Britton)

try to bring in a code of practice for manufacturers of feeds and supplements so that consumers can be assured that there are safe manufacturing procedures, quality assurance, traceability of ingredients and accurate, clear labelling which, in accordance with the law, makes no medical claims for the products. It is proposed that there will be a special logo for use on the packs of such products so that customers will know that they are reliable and safe.

I believe that horse owners and managers should also make much more stringent enquiries about the ingredients and manufacture of the products they are buying (both feeds and supplements) rather than accepting the marketing hype, because the health and future prospects of their horses, their horses' offspring and possibly their own livelihoods could all be at stake. Having done this myself, I have been surprised at how many products do not stand up to rigorous examination but, conversely, how many unexpectedly turn out to be really worthwhile.

What's in a food?

Nutritionally, the most demanding times of a horse's life are when it is growing rapidly and during lactation. The foetus makes heavy nutritional demands on the mare during the last three months of pregnancy, when it is growing and developing fast, the young foal grows most rapidly during the first three months after birth and the mare needs to provide for the foal and for herself during both of these periods (a total of half her year). Together these make considerable demands on her dietary resources. The foal's intake of food other than milk during the first three months is low so the dam must provide almost all its nourishment and her own.

The nutritional constituents of horses' feed are:

- **Carbohydrates (starches and sugars)** which provide energy and heat. Amounts above the horse's immediate requirements are stored as fat in various fat depots around the body and are also stored as glycogen (the main source of energy) in the liver and muscle cells.

71

- **Fats (lipids/lipins)** which also provide energy and heat. They are energy dense, which means they provide concentrated energy – about one and a half times as much as carbohydrate – and are therefore useful for horses at the limits of their appetites.
- **Proteins** which are the nutrients used for actually making body tissue. They are therefore particularly crucial for youngstock and mares in late pregnancy. Excess protein can be stored as energy-giving fat but then loses its tissue-making abilities.
- **Vitamins, minerals and trace minerals/elements** which are needed in varying amounts but are unlikely to need supplementing in animals on an adequate and well-balanced diet composed of varied, good-quality feeds, including good grass appropriate to their type and breed. The trace minerals, in particular, are needed in tiny amounts. Much harm can be done by well-meaning owners who give supplements without first taking expert advice from a vet or nutritionist. This can adversely affect the health of mare and foal and the correct development of the latter.

Feeding the right amount

The best general guide to feeding equines is to do so according to bodyweight and the animal's 'job', in our case breeding. A few simple guidelines will help. A good general guide to any horse's condition is to feed it so that the ribs cannot be seen but can be felt under slight pressure. This applies to broodmares and stallions as well. In addition, the topline should be well but not excessively covered and rounded, depending on the type of animal, with the withers, spine and hips not being prominent nor, conversely, there being pads of fat on the neck, back, loins and quarters or even, as in some animals which are usually of pony, cob or heavy-horse ancestry, a channel running along the spine down which water could run. This used to be a source of pride in days gone by, but now we know better!

Animals with much Thoroughbred blood will do well on good grazing like this for much of the year. (Vanessa Britton)

An excellent way to keep track of your horses' condition is to condition score them. Thin and fat animals do not show good conception rates. It is known that both mares and stallions, particularly plump ones, on a decreasing plane of nutrition, are less fertile than slightly lean ones on a rising plane of nutrition.

There are two systems of condition scoring in common use, one more precise than the other. One rates horses on a scale of 0 (emaciated) to 6 (obese) on which a score of 3 represents correct, healthy condition as described above. The other scale runs from 1 to 10 with the same parameters, on which 6 is regarded as correct, healthy condition. This latter scale seems to be more commonly used in the USA. It is a good plan to score broodmares and stallions weekly so that action can be taken should they start to lose or gain weight.

The total weight of feed a mature horse, cob or pony (including a broodmare in early pregnancy) should receive daily can be fairly accurately arrived at by feeding according to the animal's bodyweight. The most accurate way of determining this is obviously to take the horse to a weighbridge and stand it on it wearing only a bridle or headcollar. Weighbridges are to be found in many areas where there are farms and livestock markets so it would be worth the occasional trip to your nearest one.

Once you have the accurate weight, assess the weight again at home using the following method and see if you get the same figure; any difference can then be taken into account in your future regular monitoring of your horse's weight. Measure the horse around the girth using a piece of string or binder twine. Place this just behind the withers and right around the ribcage keeping it vertical to the ground. The string should be just tight enough to press in the flesh a little and you adjust it to the appropriate tightness when the horse is breathing out. Measure the length of your twine and then gauge the horse or pony's weight by using the tables given. Weighbands or tapes which are calibrated to give you the correct weight and total feed requirement are available from various sources, including feed merchants and tack shops, but they are not as accurate.

SEE TABLES BELOW

You will notice that there is one table for cobs and ponies and a different one for horses. Cobs and ponies and their near types can be difficult to feed, as they usually have a tendency to being overweight (and even laminitic) rather than the reverse. As they evolved to thrive on relatively poor keep such as moorland and upland, improved pastureland can be dangerously high in nourishment for them. They will rarely need concentrates, particularly if their forage and roughage sources are good, and any temptation to put pony and cob broodmares on 'better' grazing because they are in foal should be

---

Table 1. Ponies and Cobs

Girth in inches	40	42.5	45	47.5	50	52.5	55	57.5
Girth in cm	101	108	114	120	127	133	140	146
Bodyweight in lb	100	172	235	296	368	430	502	562
Bodyweight in kg	45	77	104	132	164	192	234	252

Table 2. Horses

Girth in inches	55	57.5	60	62.5	65	67.5
Girth in cm	140	146	152	159	165	171
Bodyweight in lb	538	613	688	776	851	926
Bodyweight in kg	240	274	307	346	380	414

Girth in inches	70	72.5	75	77.5	80	82.5
Girth in cm	178	184	190	199	203	206
Bodyweight in lb	1014	1090	1165	1278	1328	1369
Bodyweight in kg	453	486	520	570	593	611

(Tables based on work of Glushanok, Rochlitz & Skay, 1981)

resisted. They will do best on the type of land meant for sheep or, at most, dry cows. Possibly the only extra food they may need will be a balancer, and that only on the advice of a vet or nutritionist following an assessment of their roughage and pasture sources. The sugar content of grass is lowest during late afternoon, evening and night; it starts to rise around dawn and is at its highest during the morning and early afternoon so if you only have fairly rich grazing available, it would seem sensible, therefore, to graze such animals during the evenings and at night, bringing them in as early as possible in the morning, or arranging some other routine which is suited to their needs and your circumstances.

It is impossible to tell what weight of grass your horse will eat when grazing, so if you have a pony, cob or good doer you will simply have to keep a very close eye on its weight and condition score, adjusting the time allowed for grazing and the rest of the diet accordingly. It is better to restrict the energy content of the diet rather than reduce the amount of forage. When not grazing, such animals can be kept on a surfaced area with shelter, so that they can still exercise, and have a supply of low-energy roughage (hay, feeding straw or a short-chop branded forage feed, for example) to keep their digestive systems ticking over.

Many Thoroughbred-cross animals, including warmbloods, may be very good doers whose weight and diets will need watching whilst, at the same time, giving them the grazing and turnout necessary for a healthy pregnancy; this same basic advice also applies to stallions. When you have a good idea of your animal's weight and have condition scored it so that you can make allowances, if necessary, for its condition, you can use the following guidelines to work out how much total weight of feed, including forage (hay/haylage/forage feeds), and concentrates if they are used, to give daily.

You can feed for maintenance (resting or very light work), work, late pregnancy/lactation or growth. (From a nutritional point of view only, a horse can be regarded as mature when over three years of age.) Mares in early to mid-pregnancy and stallions may be working and should be fed accordingly. Stallions coming up to and during the breeding season should be fed as for moderate to hard work, depending on how many mares they will have and on their individual type and constitution, including whether or not they are good doers. It should be stressed that these standard guidelines are for animals receiving little or no grass other than winter grass.

The **energy level** of a feed is measured, and stated on the analysis panel or label, in megajoules (MJ) of digestible energy (DE), and the more resources a horse is using for work, breeding or growth, the higher energy-level feeds it needs.

- For a **maintenance ration for ponies, cobs and good doers**, feed 2 per cent of bodyweight or even less if the animal is overweight. For other horses, feed 2.5 per cent. Energy level: 8 to 8.5 MJ of DE, less for fat animals.
- For **moderate work**, including riding/Pony Club work, showing, half a day's hunting at weekends, manège work, hacking with some cantering and jumping, feed 2.5 per cent of bodyweight but perhaps use higher-energy feeds (see below). Energy level: 9 or 10 MJ of DE.
- For **hard work**, including high-level competition work, hunting three full days a fortnight etc., feed at 3 per cent of bodyweight with a fairly high-energy feed. Energy level: 12 MJ of DE.
- For **Weaned youngstock and lactating/late-pregnancy broodmares**, a maintenance ration would be 3 per cent of bodyweight daily. Energy level: 12 MJ of DE.

Although **protein** is obviously important in the diet, especially for breeding stock, too much can cause problems in any category of animal. Protein has to be metabolised, and that uses up energy. Too much can cause lethargy, excessive sweating, kidney stress (as they try to filter out the excess protein metabolites or waste products) and apparent general discomfort in the horse, including raised pulse and respiration rates. Horses generally need less protein than we would imagine: mature horses in hard work only need 8.5 per cent protein in the total diet. The categories of animal needing more include breeding stock, old horses and sick and debilitated ones.

The protein content (expressed as crude protein) will be stated on bagged, branded feeds, whether they are fibre feeds such as bagged haylages or concentrate feeds. They all come in various grades of energy and protein to accommodate the needs of just about every type of equine, doing every job you can think of, including the various categories of breeding stock.

As examples, average-quality hay will contain about 6 or 7 per cent protein and good grassland can contain about 20 per cent protein. Well-made farm-produced haylage (in plastic-wrapped round bales) can often contain a percentage protein in the teens and with energy levels of around 10 to 12 MJ of DE. An analysis (which a good supplier will probably have had done) is a big help!

- **Foals up to three months** need most protein in the total diet at 18 to 20 percent.
- **Weanlings at six or seven months of age** need 16 per cent, gradually reducing to 14 per cent.
- **Yearlings** need 13.5 per cent.
- **Two-year-olds** need 10 per cent, although up to about 12 per cent will be needed if they are flat-racing Thoroughbreds.
- **Mares in the last three months of pregnancy** need 11 per cent.
- **Mares during the first three months of lactation** need 14 per cent.
- **Mares from the fourth month of lactation to weaning** need 12 per cent.

Because of the variability of grassland, it is a good plan to have your soil and herbage assessed by someone who understands the needs of different categories of equines: the various fertiliser companies will often do tests free of charge if you are buying their products and you can always contact the Equine Services Department of the Agricultural Development Advisory Service (ADAS). The results can be discussed with your vet or nutritionist and taken into account when formulating your horses' entire diets, including grass, hay or other fibre feed and concentrates.

Because horses evolved as fibre eaters (grass and other plants which, in domesticity, includes feeding straws such as oat and barley, hay, haylage and nowadays branded forage feeds, usually short-chopped and based on straw and dried alfalfa or lucerne), it is sensible, as confirmed by all modern research, to make fibre, not cereal concentrates, their staple diet. Even horses performing strenuous work are now known to need no less than 50 per cent fibre in their diets for optimal digestive function. Moderately working horses do well on 60 to 75 per cent fibre and those on maintenance rations 75 to 100 per cent fibre.

However, mares in late pregnancy and foals are an exception, and will need both **fibre** and **concentrates**.

- **Mares in late pregnancy** will probably do best on 50 per cent fibre (including grass) and 50 per cent concentrates although ponies, cobs and good doers will need less, if any. As the foetus takes up more and more internal space, the mare will simply not have the room for large amounts of fibre and will probably not want it. A good branded stud mix or coarse mix (sweet feed) or cubes (pellets) will provide her remaining needs at this demanding time, when the foetus is taking more and more nourishment and, as foaling approaches, she gears up for lactation.
- **Foals** cannot digest fibre for some months as it takes time for their digestive systems to develop physically for this process and for them to build up the necessary complement of micro-organisms in the large intestine without which no horse, young or old, can digest the cellulose plant cell walls of fibrous feeds. Foals are sometimes given a course of probiotics to help with this process and they eat their dams' droppings for the same purpose as they will contain such micro-organisms (yet another reason for meticulous parasite control).

Foals should live on the 10 to 20 litres (17½ to 35pt) of milk (depending on whether they are ponies or horses) produced by their dams during their first two or three months to provide the correct growth. If the mare's milk supply is inadequate, adjust her diet until the milk comes up to standard. It is better to feed the mare to alter the milk's nutrient content and quality rather than to supplement the foal's diet with concentrates at this stage. If the foal is doing well, there is no need to introduce concentrates until about eight weeks before weaning. This will help to adapt its digestive tract to concentrates in preparation for weaning and its first winter.

If you have the opposite problem of a foal growing too fast (which can be disastrous for the strength, conformation and function of its mature skeleton), your vet or nutritionist may well advise that all concentrates for mare and foal must be cut out so that the nutritional content of the milk is reduced and the foal will not have access to 'extras'. Some mares with very generous milk amounts and qualities do so well by their foals that over-rapid growth is a problem and in extreme cases you may be advised to muzzle the foal for short periods each day.

Concentrate rations formulated for feeding foals are usually fairly low in energy but high in protein, vitamins and minerals. This is because mare's milk is usually high enough in energy for a foal's needs. Concentrate rations usually contain more energy and protein than fibre rations. It is not always easy for owners to work out the precise amounts of each which will make up the total daily requirements of the animals in their charge, even with the information available from pasture, concentrate and fibre analysis. This is why it is always a good plan to enlist the help of a vet or nutritionist to ensure that the animals' all-important feed intakes are suitable for their needs.

We hear a good deal about the vital **calcium**: **phosphorus** ratio of equine diets and recommendations have regularly changed over the years as more research knowledge becomes available. There should always be more calcium than phosphorus in horses' diets for healthy bones and teeth; the ratio, and amount, of each, plus **magnesium** – which is the third mineral of which bone is composed – is particularly important in foals' diets. Bone contains a calcium:phosphorus ratio of 2:1 and it is normally recommended that the total diet should contain a slightly greater ratio than this because not all the calcium is digestible by or available to the horse. The presence and amounts of other minerals and substances can also affect the absorption of calcium. However, too much calcium can hamper the absorption of other important minerals such as iron, manganese and magnesium.

Foals' concentrate rations need to be nutrient-dense with more minerals than those of the dam to ensure adequate bone mineralisation. It is possible to buy foal concentrate rations which supply the correct amount and balance of minerals without overdoing the energy, to help avoid excessive growth.

The dam must take care of her own mineral needs as well as supplying a different balance for her foal in her milk. Since minerals (and also vitamins and trace nutrients) must be properly balanced within often fairly narrow limits, because of the effects each has on the other, I would again recommend that you take expert advice on suitable rations rather than, for example, adding, indiscriminate amounts of the various calcium supplements because you know that foals 'need plenty of calcium'. The advice is usually free these days so there is no reason not to ask for and follow it.

Quality of feed

It is never worthwhile to skimp on feed quality for either mature or growing breeding stock. Poor-quality feeds are not only low in nutrients, often poorly balanced and somewhat indigestible, but they may also contain disease- or allergy-causing pathogens and other toxic substances. Use the best-quality feed available and feed it in accordance with expert advice.

Remember that the feed content and digestibility of grass and other plants will vary throughout the year. Spring grass is the most nutritious and highest in sugars and proteins, there is a dip in quality in midsummer and then a carbohydrate-rich flush in the autumn. These factors all have to be taken into account when feeding concentrates and any supplements. Seasonal pasture analysis is a good idea, so that any adjustments can be made to the diet by means of appropriate concentrates, supplements or balancers.

Although ponies, cobs and good doers can usually manage very well on reasonable-quality grass, they cannot live and breed on thin air and a mouthful of soil, and will not do well on poorly balanced feed. So, for any category of animal, it is worth keeping an eye on the nutrient content of their diets as well as their individual body condition.

Chapter 9
Diseases and Abnormalities

This chapter may have a pessimistic air about it but just remember that most pregnancies and foalings go off without problems, particularly if the management is good. After birth, too, proper care will normally ensure the health and wellbeing of both mare and foal. However, it is always a good idea to know what is most likely to go wrong if anything does, how to recognise when you have a problem, and to be willing to call in veterinary help quickly.

Laminitis in pregnancy

We hear so much about ensuring that pregnant mares receive a correct and balanced diet with plenty of nutrients to provide for themselves and their foetuses, so it can be a real blow to learn that a pregnant mare has gone down with laminitis.

Research continues on this devastating and extremely painful disease. It is still not fully understood, although everyone seems to agree that it is caused by a disturbance in the blood supply to the feet, perhaps by a chemical change in the blood and also sometimes by trauma to the feet such as excessive weight-bearing and concussion. It is normally pony and cob mares which contract the disease, although others can also do so – for instance, lack of foot care in Thoroughbred broodmares is the commonest cause or trigger factor for laminitis.

Difficulties can arise in feeding pregnant mares as laminitics need to be on a low plane of nutrition, yet with specific nutrients (they must definitely not be starved as this can actually encourage the condition or make it worse if it is already present). Mares with chronic laminitis or founder should certainly not be put in foal.

Prevention through a correct diet, good foot care and judicious exercise is always the best course of action with susceptible animals. Pregnant mares can be hacked out until two or three months before foaling, although it would be better to lead them briskly in hand during the latter period. Careful exercise is good for them

and better than being confined to their stables or a small dirt yard for fear of their getting laminitis – which some do at a whiff of grass. Even in the field, many pregnant mares laze about too much and get thoroughly out of condition.

An alfalfa- or lucerne-based, short-chopped fibre feed is an excellent choice for mares susceptible to or with laminitis, plus a special supplement formulated for hoof health. The safest time for them to be turned onto pasture for a couple of hours of grass is evening. If the mare actually loses too much condition, add soaked sugar beet pulp to the forage feed or look for one of the branded feeds specially formulated for laminitics. You may also be advised to feed a tablespoonful per day of potassium chloride in the feed, as potassium seems to have a beneficial effect in maintaining the integrity of the soft tissues in the feet and on the circulation to the laminae. Roots and succulents will make the diet more appealing.

Foaling difficulties (dystocia)

Because mares give birth more quickly than most other domestic mammals, things can go wrong more quickly and you need to keep a watchful eye out (if you can catch the process at all). Chapters 6 and 17 deal with the normal foaling procedure, but there are three words about which it helps to be clear when discussing foaling:

- **Position**: This means the position of the foal's spine in relation to the dam's during foaling. The foetus is normally born with its spine uppermost (called the dorsal position) and turns into this position during first-stage labour, having spent the last third of the pregnancy in the ventral position (chest uppermost towards the dam's spine).
- **Posture**: This means the attitude of the foetus's neck and legs in the uterus and during birth, for example whether they are flexed or, in readiness for birth, extended.
- **Presentation**: This is the way in which the foetus

is presented to the birth canal for foaling. An anterior (head-first) presentation is normal. A posterior presentation means the hindquarters are coming first (which is abnormal) and a transverse presentation means the foetus is lying across the birth canal (again obviously abnormal).

If there are going to be problems at foaling, probably one of the most common is a postural one in which one of the forelegs, or maybe the head, is turned backwards. This blocks the foal's smooth passage through the birth canal. Like all problems, this must be detected and corrected

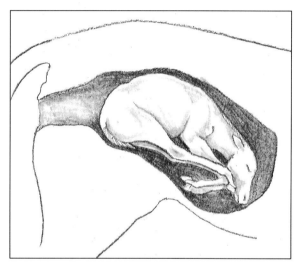

If the foetus is presented tail-first there is no chance of the mare delivering it and veterinary help must be called for.

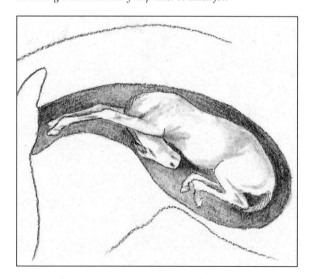

Another abnormal presentation is for the foetus to have the head bent under or to one side. Again veterinary help will probably be needed to correct this although an experienced stud groom could probably do so.

quickly otherwise the matter will become much more serious. If the mare is straining with no result, abnormal posture may be the cause. An experienced stud groom can usually deal with it by inserting a lubricated hand and arm into the canal and correctly positioning the leg or head, after which delivery is usually rapid and normal. Many amateur breeders will be their own 'stud grooms', so the vet should be called in the case of any abnormality. If the problem is a malpresentation of the foal, an abnormal position or some other serious problem such as a deformed, diseased or very large foal, or a problem with the mare herself such as not straining to deliver the foal maybe due to internal injury or weakness, veterinary attention will also certainly be needed.

The main signs of trouble are:

- the mare having 'broken water' but with no sign of delivery after about fifteen minutes
- anything less than two hooves and a muzzle showing or no progress in foaling
- if the placenta has separated from the uterus and is coming with the foal; instead of the normal white membrane, there will be thicker, red membranes which prevent birth occurring normally, and to prevent the foal suffocating, the placenta must be cut open and the vet called at once.

Most problems of foaling are caused by a diseased, deformed or dead foetus.

If the vet is unable to correct a problem fairly simply, the mare may need to be anaesthetised so that he or she can work more easily and, if necessary, perform a Caesarean section, which is a major operation. Depending on circumstances, the operation may be performed on site or it may be necessary to take the mare to a veterinary hospital.

The mare may subsequently be given intravenous fluids to counteract dehydration and shock, antibiotics to fight any infection and possibly oxytocin to encourage the contraction of the smooth uterine muscle to pass out fluid and other debris from the uterus. The vet may administer fluids into the uterus for the same purpose.

After any major operation or a difficult birth, the mare may suffer from severe colic, internal bleeding or laminitis (due to disturbance of the blood chemistry or toxaemia – blood poisoning) and will need particularly careful observation for several days and, if appropriate, prompt veterinary attention. If she is not looking after

Overfeeding and/or an imbalanced diet can cause developmental abnormalities such as the 'contracted tendons' shown by this young foal which are actually the result of their being unable to keep up with the cannon bones growing too fast.
(Vanessa Britton)

her foal, appears ill, uncomfortable or disturbed, or is not eating and drinking normally, call the vet immediately.

The placenta (afterbirth) should be expelled within three or four hours otherwise uterine infection and shock can set in, which can cause toxaemia, laminitis and even death. If it is not expelled, call the vet. Do not pull out a partially expelled placenta but tie it up to itself and, when it is finally expelled, keep it in a cold place for the vet to examine.

Rejected foals

Sadly, not all mares are good mothers; according to studies in the USA and Egypt, the proportion of Arabian mares which reject their foals is slightly higher than mares of other breeds.

Rejection may take the form of the dam simply refusing to let the foal suckle because of tender teats and not being used to having her udder handled. Gentle but persistent restraint whilst the foal suckles for the first few times may overcome this, and all mares should have their udders regularly handled and washed, partly to accustom them to being handled here and partly for hygiene and parasite control.

If the mare is aggressive towards her foal, a very mild sedative or tranquilliser may be administered for a day or two to try to allow the bonding process to occur in safety. The mare and foal may also be separated by a partition which allows them to see each other, being brought together, with a handler holding and restraining the mare, only when it is time for feeding – which will be every hour or so round the clock at first, until the dam accepts the foal. If she shows no sign of accepting the foal within a very few days, it is probably best to try to find a foster dam for it or, a poor second best, to bottle-rear it as it will then, for all practical purposes, be an orphan. Dams which persistently reject their foals should not be bred from as there is no point using mares which are not going to rear their foals well and there is some evidence that the condition is inherited.

Orphan foals and bereaved dams

Few situations are sadder than a foal losing its dam or a mare her foal. Although foals can be devastated by the situation, mares which normally have the kindest natures can become so distraught that they become savage. It is essential

79

Orphan foals do much better if they have another equine to play with. It is important that they do not grow up not knowing how to socialise with other horses or ponies. (Vanessa Britton)

to enlist the help of your veterinary surgeon to help you. The National Foaling Bank may also be able to help; it can be reached on 01952 811234.

Both parties are best off if an adoption can be arranged. This will involve skinning the dead foal, draping the orphan in the skin and very carefully introducing it to the bereaved dam. The process is not easy and by no means always works – expert help is essential. The process may be assisted by administering a mild sedative or tranquilliser so that the mare will allow the foal to suckle or let herself be hand-milked.

It is normally necessary for someone to stay with the mare and foal for hours or even days and nights to prevent the mare possibly savaging and even killing the foal until they have bonded. Otherwise, they can be separated by a partition, as described above. A mare who is bereaved through having rejected, or even killed, her own foal is obviously not suitable for motherhood or adopting.

If the foal has not suckled at all and it has not been possible to milk the dam of her colostrum, the foal may need to be given colostrum (maybe by stomach tube) from other mares on the stud or in the surrounding area in an emergency. A foal should have up to 2 litres (4 pints) within six hours ideally, and before twelve hours for the effective absorption of antibodies. Milk about 50ml (1¾fl oz) from each suitable mare so that their own foals do not go without. Obtaining

colostrum in this way and freezing it at -20°C (0°F) in anticipation of emergencies is a wise precaution. It should obviously be warmed to blood heat before feeding by allowing it to thaw or running warm water over the container. Do not microwave it as this will kill the antibodies and defeat the object of feeding colostrum as opposed to other milk.

You can make a suitable feeding bottle, if you cannot obtain a proper one, by using a plastic lemonade bottle with a calf or lamb's teat on the opening; a lamb's teat will need the hole slightly enlarged. Teats for human babies are too small and therefore dangerous. Colostrum not only acts as a mild laxative to help with the passing of meconium (the foal's first dung), it is also the foal's means of acquiring immunity from the dam. Therefore, without the benefits of colostrum it will need to be given immunity in some other way. Your vet should be able to arrange for a suitable preparation to be made by taking blood from the dam or a suitable donor mare and separating the plasma for intravenous injection into the foal. Alternatively, commercially available immunoglobulin G (given by mouth) can be purchased via your veterinary surgeon, who will advise on its use.

Once the foal has had its colostrum, you will need to bottle feed it on an equine milk replacer. An agricultural replacer may be offered to you for cows, sheep or goats, but these are not correctly formulated for horses so you must stress that you want one specially for equines. Your vet may be able to provide this or you could obtain one commercially – companies which specialise in horse-health products, feeds or supplements often have milk replacers or a good pharmacy, particularly a veterinary one, should be able to get one for you. It should ideally be collected, but you can order it by mail order; if you do, stress that you want next-day delivery. Order it in plenty of time, while the foal is still drinking colostrum. Follow the instructions on the pack very precisely, including the requirement to feed the milk at body temperature (37.5°C). Keep all utensils scrupulously clean, using a baby's sterilising product or a special equine one for the purpose.

Allow the foal permanent access to water to prevent dehydration and try to get it drinking both milk and water out of buckets as soon as possible by dipping your finger in, allowing it to suck it and gradually lowering it into the liquid. Introduce warm water from a feeding bottle if necessary to get it used to the taste.

If there is any delay in obtaining the milk replacer, continue milking the mares from which you obtained the colostrum. This will only be acceptable for a very short time, however.

Initially the foal should be fed at least every thirty minutes to an hour for twenty-four hours – a newborn foal naturally suckles about every twenty minutes so if you have enough help to arrange this, it would be best. After a day or so, you can reduce the feeds to once every hour then once every two hours at a week old, increasing the amount gradually by 4–8 litres (7–14 pt) daily at around this age, depending on whether the foal is a pony or a horse foal. By the time it is seven or eight weeks old, it should be being fed every four hours and taking between 6 and 12 litres (10½–21 pt) of milk daily. Fluid intake is calculated by bodyweight so ask your vet how much liquid your foal will need daily.

You should also give the foal a salt lick and someone should regularly spend time watching it to see whether or not it is drinking water and using the lick. If not, tell your vet. Concentrate feed can be introduced at a very few days old, the object being to adapt the foal's digestive system to eating concentrates, and later fibre, as quickly as possible.

As you can see, bottle-feeding an orphan is a really onerous job and it is very much better to find a foster dam if possible, not least because of the natural equine socialising the foal will need to enable it to think and behave like a horse when it is older. Many breeders believe in allowing orphan foals to mix with other horses whilst they are still being hand-fed regularly. If you do this, it is a good idea to have a quiet and friendly 'nanny' mare in the paddock which can act as its protector, if it will.

Suckling foals can be adopted by mares at almost any age, so if at first you cannot find a foster mare, keep trying. Other animals are often used, such as sheep, goats and donkeys, but there is nothing like another pony or horse. Constant company and a view of the world are extremely important to the foal's sense of wellbeing. A humanised foal will be a misfit with other horses later in life and can be very hard to train and discipline; it may even end up being unwanted by humans and horses alike.

Loneliness can be heartbreaking for any horse, but particularly for a young foal – and especially one which had bonded with its dam before she died. Therefore, suitable company should be regarded as just as important as replacement milk. The mind and spirit must be nurtured as well as the body. If at any time your foal appears to be having problems, is uncomfortable, is not feeding, is not growing and gaining weight or looks ill, ring your vet at once.

Neonatal maladjustment syndrome

A newborn foal with this syndrome may also be called a barker, dummy, wanderer or shaker. It may suffer fits and convulsions or an inability to suckle, it may fall into a coma, it may pant and have a subnormal temperature, it may constantly chew things or simply make chewing motions with its mouth, and it will generally appear very mentally disturbed.

The syndrome is caused mainly by poor development of the foetus (possibly due to stress, infection or accident in the mare) or by birth trauma involving lack of oxygen and circulatory disturbances which cause brain damage in the foal. This in turn causes further problems which result in more nerve damage, and it becomes a vicious circle.

The condition can be treated, and if it is tackled promptly the foal may recover and become perfectly normal. On the other hand, it may worsen and die. The veterinary surgeon can give drugs to counter the muscle spasms and convulsions, fluid therapy may be given and the foal may be fed milk by stomach tube if it will not suck. Sedatives may be necessary to calm it down, oxygen inhalation can be given and, very importantly, good nursing is needed to protect and guide it and also to reassure the dam.

Septicaemia

This term (meaning 'septic blood') covers a variety of infections which may quickly affect newborn foals – usually within three days, although one condition, joint-ill, may occur within the first three months of life. The infection normally enters through the navel (hence the need for clean foaling conditions and for treating the umbilical stump with antiseptic once the cord has broken). It is most common in foals which take a long time to take their first drink of colostrum, so the routine blood testing of young foals to check on their immunity levels is important as it will warn vets and breeders of any risk.

The first signs of trouble will be failure to suckle, a high temperature and increasing lethargy, which may progress to a coma. The exact symptoms depend on the part of the body affected, as the condition covers various diseases and disorders, including sleeping foal sickness, joint-ill, meningitis, diarrhoea, pleurisy and pneumonia, although there may be other problems as well.

Treatment will consist of the administration of appropriate antibiotics plus whatever other techniques are appropriate to the specific condition which has developed.

Developmental orthopaedic disease (DOD)

This is an umbrella term covering any disease or disorder involving abnormal bone and limb formation and growth abnormalities. The cause is often nutritional (too much feed, too high an energy level or an imbalance of vitamins and minerals). However, it may also be genetic, hormonal (involving thyroid and growth hormones) or caused by the disruption of the normal blood supply to ossifying cartilage. Also implicated is trauma, particularly a period of too little followed by too much exercise such as when horses are turned out onto spring pasture after winter confinement or having been turned out onto less interesting surfaced areas. Obesity, conformational factors or over-rapid growth are also probable causes.

Treatment obviously depends on the exact condition and cause and may include correcting the diet or other management techniques, surgery, splinting the legs and the administration of anti-inflammatory drugs or hormones.

Epiphysitis involves inflammation of the epiphyseal or growth plates in the legs and can be caused by over-stress such as too much exercise; some restless dams cause their foals to follow them ceaselessly instead of settling to graze, and some foals charge about too much, which is particularly dangerous on hard ground such as occurs during a hot, dry summer. Epiphysitis can also be caused by infection and, commonly, by imbalanced or over-feeding, particularly by high energy levels. Foals which grow too quickly are very prone to epiphysitis. The affected joints will be hot and swollen, the foal will be very uncomfortable and lame and the legs may start to deform.

In the condition known as **contracted tendons**, the foal's tendons do not actually contract: rather the cannon bones grow faster than the tendons, which cannot keep up, and exert a pull on the bones in the feet to which they attach, causing the fetlocks to buckle over in extreme cases so that the foal is walking on the fronts of the fetlock joints. The legs may deform sideways (**angular limb deformity**). Milder (but still serious) cases involve pasterns becoming more and more upright and the foal appearing to stand on its toes.

Osteochondritis/osis dissecans is another disorder of immature joints due to abnormal development of bone from cartilage. There will be inflammation, pain and lameness and, upon examination, it may be found that there are small, detached pieces of bone or cartilage moving about in the joint which cause great pain.

Wobbler syndrome (ataxia), known as 'wobblers', is a condition which may appear from three months of age but rarely after three years. It is characterised by lack of co-ordination, usually in the hindquarters, although sometimes the forehand is affected. The animal usually staggers, drags its feet and is unsteady, particularly when backing, turning and stopping. Getting up is particularly difficult and those animals which cannot manage to do so usually have to be put down. Because the condition may be inherited, affected animals should not be bred from. Wobbler syndrome affects the growth plates of the vertebrae and can be due to traumatic damage to them or to abnormal development. Both can cause damage to the spinal cord, producing the above symptoms.

This limb deformity looks alarming but the fetlocks may well correct themselves within a few weeks after birth.

Retained meconium

The foal's first droppings, called meconium, should be passed soon after its first bout of suckling. They are stored in the large intestine before birth and the colostrum's slightly laxative quality helps in their passage. The meconium is usually dark in colour – brown, green or blackish – and in pellet form.

If the foal does not start to pass meconium soon after birth – and particularly if it appears to be straining to do so, maybe even rolling, appearing colicky, lying on its back or in other unusual postures which indicate pain and discomfort, refusing to suckle and with a distended abdomen due to gas in the intestines – the vet should be called at once. The vet will probably give painkillers by injection with an enema and possibly also liquid paraffin by stomach tube.

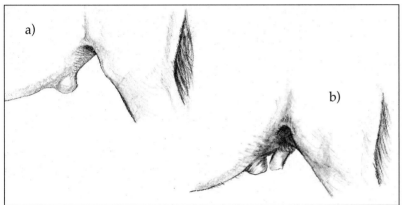

a) *An umbilical hernia in a yearling filly, and (b) in a yearling colt. Many such hernias, which occur usually during the first six weeks, right themselves during the first year and are often left to do so unless the abdominal contents become constricted and strangulated. Veterinary advice should certainly be taken in the case of a hernia.*

Hyperflexion or 'contracted tendon' in one foreleg to the extent that the fetlock is knuckling over.

Bent knees (carpal flexural deformity) may be treated by splinting the legs and applying physiotherapy.

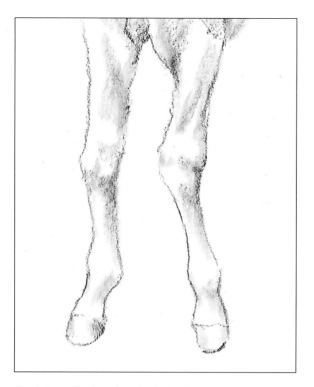

Knock knees like these (angular limb deformity) can be treated by corrective foot trimming and sometimes surgery, and the foal will need box rest with its dam.

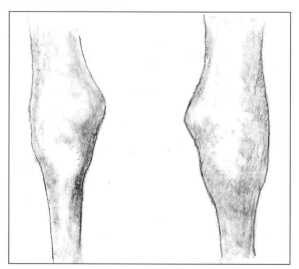

The 'lumps and bumps' on these knees are typical of epiphysitis.

After the meconium has been passed, the yellowy-orange milk faeces will start to be passed. The colour often leads novice breeders to think that the foal is jaundiced. It is normal, however, and the digestive process should now be developing and settling down.

Mastitis

This condition of the mare involves inflammation of the udder (usually only one quarter). It is painful, and the mare may prevent the foal suckling or even attack it because of the pain. It is caused by bacterial infection and can be treated by stripping the udder (milking off all the milk) and inserting antibiotics into the affected quarter. The foal will have to be fed on a milk substitute meanwhile (see 'Orphaned foals and bereaved dams') and prevented from suckling.

The udder becomes swollen, hard, hot and painful and the lymphatic vessels along the underside of the abdomen become congested. They stand out like ropes along each side, and pit when pressed. As the udder can assume a similar condition after weaning, it is important to differentiate between the two disorders.

There are other conditions of both mare and foal which may cause problems and need veterinary attention but these are probably the most common. It should be stressed that *any* abnormal behaviour, sign of ill health or simple lack of wellbeing in either mare or foal should be regarded as a potential problem and novice breeders, in particular, should not hesitate to call the vet in such cases.

The foal's body and mind must have every chance to develop in response to appropriate stress and to learn balance and co-ordination. A correct diet and plenty of liberty are the best practical ways to help this process. Here, a young Welsh Cob tests her back, hindquarters and hindlegs

Chapter 10
Artificial Insemination and Embryo Transfer

Breeding horses today often involves techniques other than a stallion and mare getting together. Artificial insemination (AI) has been the main means of producing cattle for many years, although the tide is turning somewhat as a result of the current deep depression in the agricultural industry and the cost of AI. Following on its success in this area, the method was introduced to horses with a view to producing competition animals. Embryo Transfer (ET) followed on and both techniques are now fairly common, although 'natural' methods are still by far the most usual way, in general, to breed horses and ponies.

There are many breeds of horses and ponies in the world, and many which are categorised according to discipline, not breed, such as show hunters, Western performance categories, eventing, riding horses, working ponies, dressage horses etc., all controlled and administered by their own society or association, and not all administering bodies allow the registration of animals produced by means of AI or ET. The arguments for and against the use of these techniques are many and varied and cannot be discussed in detail here. The main objection to the use of AI always used to be that there could be doubt about parentage, but this never really held water once equine blood-typing became established and it is now completely untenable because of the advent of DNA identity testing. Another objection to it was economic: because a sample of semen collected for use in AI can be divided many times, the market value of that stallion's services would inevitably plummet as its scarcity value decreased. There would also be the possibility of the population becoming saturated with certain horses' genes, which would ultimately lead to a dangerously high level of inbreeding.

All these arguments can be overcome with a little common sense and co-operation, however, as is shown by the number of sports animals successfully produced by both techniques. These qualities are not always dominant in people's thinking, however, and certain categories of equine must therefore still be produced by natural methods if they are to be registered. For example, Thoroughbred foals resulting from AI or ET may not be registered in the General Stud Book (GSB – the worldwide stud book of the Thoroughbred breed to which all Thoroughbreds must trace). They are also only eligible to race if both they and their sires and dams are the result of natural covering or service. Arab horses in the UK are eligible for registration if they are produced by AI but not ET. AI is not permitted in Italy; the semen of any breed may be exported from Argentina but not imported, while the reverse holds in other countries: animals conceived by AI may not be registered in some countries but may be registered if conceived abroad and imported *in utero*, and so it goes on.

The only advice we can give breeders is to minutely check the rules and regulations, not only of the breed society or other administration organisation with which they may wish to register their stock, but also of their national agricultural organisation (in the UK the Ministry of Agriculture, Fisheries and Foods), European Union rules and regulations if applicable, and those that apply around the world. Otherwise, one could be left with a beautiful foal worth very little in the twenty-first-century market because it is ineligible for registration.

Artificial insemination

AI involves the placing of a stallion's semen into a mare's uterus without the two ever having to meet. The technique involves a great deal of meticulous preparation if it is to have any chance of success. Because fertility rates from AI in horses are roughly 75 per cent or less (lower than those for natural methods), the stallion and mare chosen should themselves be fertile individuals, correctly fed and in a relaxed, let-down (i.e. not athletically fit) condition, happy and healthy. Any other situation may well result in failure, and also in the waste of a good deal of money

85

because, owing to the technical expertise involved and the use of specialist equipment and facilities, AI is generally more expensive than conventional mating.

AI is ideal for mares who travel badly, or if the stallion you want is well out of your area (especially abroad). It also decreases the spread of infection and the risk of injury to mare and stallion. The health and fertility tests involved for both mare and stallion ensure the best possible chances of conception while at the same time, avoiding over-use of young or old stallions in particular, because the semen can be split into many doses. Mares and stallions which, perhaps because of athletic injuries, cannot mate via the normal mounting method can still be used for breeding.

A disadvantage is the expense, although mare owners must offset against this the fact that they do not have to pay travelling and keep expenses. Unscrupulous or incompetent operators may also cause problems by not using properly sterilised equipment, not keeping proper checks on parentage or implementing handling controls to ensure that the right semen reaches the right mare. Also, unproven and young stallions may be used to bolster the stallion centre's stocks and too small a number of stallions may be used, reducing the gene pool. Careful discussion with experts in the field and, as ever, with your veterinary surgeon will help you decide whether or not AI is for you and your mare or mares.

To prepare the mare she will first need a thorough veterinary examination during oestrus to check the condition of her breeding organs. She can also be swabbed for contagious equine metritis at the same time, because although she will obviously not be able to infect the stallion, any infection will greatly reduce her chances of conceiving and maintaining a pregnancy.

The vet will scan the mare or examine her rectally, checking her ovaries and determining the presence or absence of developing follicles. This will also help decide whether or not the mare will need treatment of any kind, including the administration of hormones. Assuming a developing follicle is present in one of the ovaries, you will have to wait a few days for ovulation to occur. The mare will then be given a synthetic version of the hormone progesterone in her feed, even though she herself will then be producing progesterone from the corpus luteum which now occupies the former follicle. The progesterone will stop her from coming into season again and prevent the development of

more follicles. On the tenth day of being given the synthetic progesterone, the vet will give the mare another hormone, prostaglandin, by injection to 'kill off' the corpus luteum. The cessation of progesterone treatment kick starts the development of another follicle and four days later the mare will have an injection of the hormone chorionic gonadotrophin to stimulate ovulation. The mare will be inseminated thirty-six to forty-eight hours after the administration of chorionic gonadotrophin.

If, during the initial examination, there is no developing follicle, the mare will be given prostaglandins by injection to remove the corpus luteum and permit a follicle to develop. After ten days, synthetic progesterone is given, and events take place as described.

Not all mares are given hormones, however; some are allowed to ovulate naturally.

To collect semen, the stallion is encouraged and trained to mount a mare or an artificial dummy mare and to ejaculate into an artificial vagina (a double-skinned container with blood-heat liquid between the layers) which, in turn, is subsequently placed in an insulated container to maintain the temperature of the semen. Sometimes a condom may be used. Stallions sometimes need an in-season mare nearby to get them interested, the penis being deflected at the appropriate moment into the artificial vagina. Stallions which are unable to mount can be encouraged to 'draw' (produce an erection) and ejaculate without actually mounting simply by having an in-season mare at hand. The ejaculate can be used in its entirety or be divided into about ten or twenty parts.

The ejaculate can be used immediately or within a few hours. In the latter case, it is kept at blood heat and may be mixed with a special protective, nourishing fluid called an extender, which often contains antibiotics and nutrients to maintain the health and activity of the sperm. Semen can be chilled or frozen to extend its life; it is transported to a laboratory for screening and quality control, partly to check for infection but also because not all stallions' semen chills or freezes well; it can result in a low sperm survival rate. Both freezing and thawing distress the sperm and samples of semen with a high percentage of dead sperm will mean that the donor stallion is not suitable for this procedure. Although AI works well in cattle, its viability in equines is very variable and we do not yet know why some stallions' semen sails through the procedure while that of others is adversely affected.

There are special stallion centres in most countries now with the specialised, highly skilled staff needed for successfully carrying out AI. The stallions go there at livery and usually have to have several health and blood tests which must be certified and passed by a veterinary surgeon before they are allowed onto the centre's premises. Once collected, the semen is chilled or frozen at the centre, assessed for suitability and split into a given number of doses before being stored or sent out to mare owners or their vets in containers called straws, within special freezer flasks, by a next-day delivery service. Whereas chilled semen has a life of only about thirty-six hours, frozen semen can be kept indefinitely, but fertility is slightly lower with frozen than with chilled semen.

Not all vets are experienced in AI techniques, so some stallion centres have local and regional contacts with practices who are able and willing to carry them out. Some will send out their own staff to inseminate mares at home.

As we have seen, the vet will probably have been monitoring the state of the mare's ovaries and follicles and will contact the centre when he or she judges that ovulation is likely within a very few hours, giving instructions as to where to send the semen. Ideally, the mare should be inseminated as soon as ovulation has occurred, or just before.

The equipment used for both collecting semen and inseminating the mare must be sterile. To transfer the semen into the mare, the vet will wear a lubricated glove and hold a catheter or plastic tube to which is attached a syringe into which the semen will have been placed. He or she inserts the tube into the vagina, checks with a finger that the cervix is relaxed (open) and carefully inserts the end of the tube through the cervix into the uterus. The syringe is then depressed and the semen pushed into the uterus, after which the equipment is removed and, it is hoped, the mare will be scanned pregnant in about eighteen days' time.

Embryo transfer

This is another technique which is well proven in agriculture but which has only relatively recently been introduced to the breeding of equines. It involves removing an egg which has been fertilised by AI or naturally from a mare who normally has athletic abilities or genetic qualities which make her line valuable, and inserting it into another mare's uterus for development and foaling.

The donor mare can continue a competitive career with only slight interruption to her programme (she will be out of action for a couple of months as she must be let-down, relaxed and happy before she can conceive), building a reputation for herself and her family and offspring, without having to take years off for motherhood. If she pursued a full competitive career, depending on her discipline, she might be too old to start breeding by the time she retired. This is the main reason for using embryo transfer, and with increasing stress being placed on the performance records of both mares and stallions, the technique is likely to be used more and more.

Other factors in favour of ET are that it enables mares who can conceive but not maintain a pregnancy to continue to be bred from, and also those which are known to have difficult foalings. However, if any problematic conditions are genetic and not acquired, it should be remembered that these faults may well be inherited by the offspring and, as is always the case with any inherited faults, the resulting equine population will gradually become more and more adversely affected. Moreover, ET requires great skill and attention to detail on the part of the operators.

The preferred age of the recipient mare is from three to about thirteen years of age. In order to give the transferred embryo the best chance of implanting into her uterus and developing, the uterus should be allowed a couple of months to recover, contract and regain tone following foaling; it follows, therefore, that this extra time will also give the foal time to gain strength and maturity before accompanying the dam to the ET clinic.

The mare will also need to have all the usual veterinary checks and be swabbed and declared free from infection before the procedure can be undertaken. Ideal recipient mares are often of the roomy, half-bred type, which are known to be easy breeders with no problems cycling normally, to maintain pregnancies without problems, to be good mothers and likely to set a good example to the foal, being easy to handle and having a calm, sensible temperament with no vices. Two or three mares may be chosen to multiply the chances of success. They should all be about the same height as the donor mare and be in good general health and condition.

The recipient and the donor must have their

oestrus cycles synchronised as closely as possible, ideally ovulating within twenty-four to forty-eight hours of each other so that their reproductive tracts are all in the same state and able to provide a similar environment for the embryo. Sometimes it is possible to collect two fertilised eggs from the donor mare, so extra mares to take them increase the chances of success – and maybe the number of resultant foals. Although your registering organisation may accept only one registration per mare per year, many animals are eligible for more than one registry so both foals would have papers of some sort to add to their value.

Hormonal treatment can be given to synchronise the mares' oestrus cycles. Once synchronisation has been achieved, all the mares involved should be taken to the ET clinic before they are expected in oestrus.

The donor mare can be artificially inseminated or mated conventionally. The embryo is flushed out six or, ideally, seven days after ovulation and fertilisation, by means of a saline solution. By the eighth day it will be too big to remove. The process is repeated three or four times and the fluid left standing in special bottles for the embryo, if it is present, to settle to the bottom of whichever one it is in. Most of the fluid is siphoned off, and the remainder poured carefully into a special dish and placed under a microscope so that the embryo can be searched for, washed and sucked up into a straw (a fine tube) for transfer to the recipient mare.

The embryo can be transferred non-surgically or surgically. The former method has a slightly lower success rate but the latter involves the mare in an operation under general anaesthetic. In the non-surgical technique, the embryo is transferred by means of an insemination catheter via the reproductive tract into the uterus. It is ejected when a plunger on the other end of the catheter is depressed. As I mentioned, there is slightly more chance of the embryo being damaged with this technique.

In the surgical method, the recipient mare has the embryo inserted directly into her uterus. An incision is made just above the udder and one horn of the uterus is brought out through it, a small hole is made in the horn, the embryo is injected from its straw into it, the horn is replaced and the incision is stitched up.

A different surgical method involves the mare being given a sedative and local anaesthetic (being restrained in examination stocks) and the incision being made in her side just in front of the wing of the pelvis, through which the embryo can be placed into the appropriate horn of the uterus. In ten to fourteen days, she can be scanned to see if the procedure has 'taken' and she is pregnant.

PART 2
PRACTICAL ASPECTS

Chapter 11
Decision Time

Why do you want to breed a horse? As this book is about basic horse breeding, I presume that most of its readers will be novice breeders (although not, I hope, novice horse people because the two together add up to a very dangerous situation). You may well then, be breeding a horse or pony for the first time. Almost certainly you will not be standing your own stallion, which is why the emphasis throughout the book is on mares and youngstock, with stallions receiving less attention.

You will have probably seen many articles urging you to think carefully before getting your mare pregnant, pointing out all the pitfalls, problems and expense (even when things do not go wrong), telling you in graphic detail about all those things which *can* go wrong, and finally reminding you that it is almost impossible to make a decent profit out of breeding equines. Indeed, the preceding section on veterinary considerations may have put you off already. If you are still set on your course, given all that, the question is pertinent: why do you want to breed a horse?

The wrong reasons

If you want to breed for any of the following reasons, think again:

- You want to make a lot of money or get rich quick.
- You know you can breed a sure-fire world champion.
- Your mare has become unsound for some reason (not an accident) and breeding from her means she can still be useful and earn her keep.
- Your mare is extremely difficult to handle, ride or drive, bad-tempered, unco-operative, very highly strung, etc. but if you send her to a gentle stallion, he will correct all that in the foal.

- Your mare has a poor conformation and constitution, but the stallion can correct all that in the foal.
- Someone else, quite experienced, failed to make a successful broodmare out of your mare, but you feel you can do better.

There are various other reasons not to breed horses or ponies, but the above are perhaps the most common. Let us take them one by one.

It is extremely unlikely that anyone will make a lot of money out of breeding horses unless they already have a lot of money to put into their stud and use only top-class mares, stallions and human expertise. Even then, the wastage in breeding any animal for a specific performance purpose is high.

Breeding a sure-fire world champion is impossible. Despite the most assiduous checking of conformation, performance history (competitive, hunting field or otherwise), temperament and genes, you can never know just what your mare and her mate will throw. Genes can do funny things and skip several generations, particularly in unknown combinations. Breeding is always a gamble, even though you can shorten the odds somewhat.

Never breed from any mare or stallion which is unsound for any reason other than a genuine accident. This may seem like a counsel of perfection but, as crocks tend to produce crocks, it is your only safe way of being even half sure that you are going to get something worthwhile. It is true that some unsoundnesses are brought on by bad management (e.g. wind and foot problems) or ridiculous amounts of overwork, perhaps leading to fatigue or accidents and ultimately degenerative joint disease (DJD), but you will rarely be in a position to check on the historical reason for unsoundness. To be safe, use only sound animals. Broodmares and stallions

Whatever type you want to breed, whether it's noble heavy horses . . . (Vanessa Britton)

should not be regarded as throw-out equines which are no good for anything else – not if you want to breed good stock.

It is a bad idea to breed from any animal with a suspect temperament, even though it may have deteriorated due to bad treatment. Tricky animals need experts to handle and work them, and you cannot guarantee that your foals will always end up with experts. The matter of heredity should also be considered: there is no guarantee that a kind stallion's genes will dominate those of a bad-tempered mare – you are just as likely to get a bad-tempered foal as a kind one.

Equally bad is to breed from a mare which has poor conformation, a weak constitution, poor feet or crooked legs. However beautiful (and expensive) the stallion may be, he is not a miracle worker. You are unlikely to end up with a foal which is just like him; you will most likely get a cross between his good points and the mare's bad points, which you do not want. A horse is an athletic animal and should be bred for hard work, unless you only want to keep it for yourself and potter around the lanes. Otherwise you must have good, sound conformation and constitution if the foal which you are responsible for bringing into the world is to have any chance of being wanted and leading a useful life.

Breeding from animals which are poor breeders – which are infertile, are prone to pregnancy problems, have poor health, have a

. . . or quality riding stock, have it clear in your mind what you are aiming to produce and what market there will be for your youngstock. (Vanessa Britton)

poor milk supply and have little or no maternal instinct – is sheer folly. You cannot give a foal a good start in life with a dam like that. If someone with experience and a feel for horses has cast off your mare, there is a very good reason, one which you are most unlikely to be able to overcome.

The right approach

On the other hand, if your approach is along the following lines, it may be worth giving it a try.

- You love *good* horses.
- You can afford to run at a loss, at least initially.
- You accept that breeding is always a gamble.

- You prefer to have a few or just one high-class, sound mare of good temperament rather than several lesser ones. Ideally, she will be a known good breeder and have a proven performance record.
- You are prepared to go to, and spend your money on, a really high-class stallion with proven performance and stud records and good fertility rates whose conformation complements that of your mare and who has a good temperament.
- You are sure there is a market for the type of foal you plan to breed (unless you can be sure of being able to keep him yourself).
- You have the money and facilities (land, stabling etc.) to do the job properly.

91

- You are willing to learn as much as you can and also call in expert help.

Most (though not all) people who are interested in horses love them. It is admirable, laudable and understandable to buy a horse because you feel sorry for him or her. However, all the hard-headed experts say that buying a horse because you feel sorry for it whilst convincing yourself that you can always breed from it if things do not work out is completely the wrong reason to start breeding.

The pet-food industry thrives on the poor horses, which usually end up in its tins long before the end of their normal lifespans. Poor horses (poor conformation or bad temperaments) rarely give satisfactory service or last long physically. There are plenty of people breeding poor horses and ponies, and there are plenty of domestic and feral foals which never even see their first birthday, let alone a saddle, bridle or harness. So make your motto 'The best, not the rest' and aim for excellence – not perfection because it does not exist in horses and you will be forever dissatisfied. There is always a market for really good animals – even if it is 'only' for breeding!

Any new business will probably run at a loss for a few years, so you must be prepared for that lean time and financially able to support it – ideally not from loans or overdrafts, in case things do not work out. And although breeding is a gamble, you can lessen the odds against you by using only good, sound animals, having facilities which are at least adequate, having more than enough cash available, and being prepared to admit that you do not know everything and that expert help rarely goes amiss. Moreover, going for quality, not quantity, will tilt the odds in your favour. A good stallion on a good mare will virtually guarantee a good foal. If one partner is substandard the equation will not work.

Breeding for yourself

Some years ago, there lived in England a Dutch gentleman who excelled at many pursuits, including breeding, schooling and riding horses (also writing books about them, including *Equitation, Dressage* and *Horse Breeding and Stud Management*). His name was Henry Wynmalen and in the latter book he went against all common advice and suggested that one-mare owners should breed from their much-loved but maybe unremarkable animal because, he reasoned, she must have some good qualities for her owner to love her so much and to want to keep her. There is something to be said for that argument. The trick lies in sorting out the various elements in such a mare (because it is notoriously difficult to judge objectively an animal you love) and in sending her to a stallion which is particularly strong where she is weak but is also *not* weak where she is strong.

If you are breeding from your mare because you want another horse for yourself (not to sell) but especially because you want something of her when she dies, I can only say go ahead and continue her line – but the best advice is still to choose the very best stallion you can find, almost regardless of where he is. Certainly never choose a stallion mainly because he is cheap and close by.

Breeding to start a line

If you want to breed to start a successful line of horses or ponies in whatever equestrian discipline you choose, it makes sense to start with animals which are already proven performers and, ideally, breeders in that field. If a mare you own already does not have those qualifications but could have had if you or her other owners had followed that route, by sending her to a stallion who is himself proven in that discipline you will be on the right track.

On the other hand, if you own an unremarkable pony, cob or other sort of mare and intend to breed world-class eventers, you need time, money, patience and a lot of luck on your side. This is because, although it is true that most of the best eventers are about seven-eighths Thoroughbred with a dash of native or 'heavy', and your native-type mare could provide that final eighth, you will have to wait many years to produce the three generations of Thoroughbred infusions necessary before you reach your pedigree goal. Each foal will need to be a filly (unlikely) and, to be fair and sensible, you will have to wait until each filly is four years old before putting her to a suitable Thoroughbred. You will, then increase the risk because you will not be breeding from proven performance females, as they will be too young to perform. You will finally need another eight years or so after the birth of the third offspring before you can watch it gallop round Badminton – if it qualifies. This procedure would take twenty-three years, allowing for pregnancy and maturing, but only if you get fillies each time.

You need plenty of space and land if you are to give your foals the facility to develop fully. Essential antics like this, which strengthen young bodies and develop agility, are repressed when animals are stabled and/or turned loose in small paddocks, manèges or indoor exercise areas. (Vanessa Britton)

There is no guarantee of that nor even of getting a foal each year.

It could be that you have such a line in progress already, in which case you could start to refine the process by using proven event stallions in future. Otherwise it would be wiser, much quicker and far less risky to invest in the purchase of a proven eventer broodmare and, if you wish, use your native mare for producing much-needed native purebreds or good children's cross-bred riding and performance ponies, using proven stallions of that type.

Breeding to start a line can be very rewarding provided you get your priorities right and refuse to use suspect animals (those suspected of poor constitution, conformation, temperament and athletic and breeding performance), however appealing their genes may appear. As I said in chapter 4, inbreeding (the mating together of close relatives) is very risky for novices (or even experts) but line-breeding (mating together more distant relatives) not only fixes good qualities but gives more certainty that the offspring will be more or less what you are aiming at. Remember

93

that bad qualities can also be fixed, so if a horse shows, or throws, bad qualities, avoid it like the plague if you want good stock with a good reputation. It is perfectly possible to breed out bad qualities by not using animals which show, or are known to produce them, and with the future advances in gene technology, it should soon be possible to screen out such undesirable elements from your stock.

Breeding to sell

If you specifically want to breed to sell, and most people who breed more than one or two animals do wish to sell, then you really do have to use the very best foundation stock you can afford. Furthermore, no animal is worth much today if it does not have papers. A pedigree alone is little use as most buyers will want to know why the animal has no other papers and is not registered. You are unlikely to be able to sell for a decent price an animal which has not passed its breed inspection or grading.

Not many years ago, competition animals with no papers were the norm in the UK and magazine articles abounded urging breeders to put their animals forward for inspection and grading and riders to keep proper records of

Foals are intensely curious and can find all sorts of objects and activities to stimulate their young minds. Here, a horse-ball has been provided for the youngsters to play with. (Vanessa Britton)

their animals' achievements, registering unregistered animals wherever they could. Now things are different. With the advent and acceptance of the undoubted abilities of continental European warmbloods, particularly for showjumping and dressage, commercial breeders know that papers are essential if they want a good price, or even a sale, for their stock.

In other fields, breed societies have always demanded registered animals for their approved classes, qualifiers and so on, but now more and more pure show animals are also registered in some performance register or other, particularly mares which may later be used for breeding. As there are registries for animals of no specific breed but of good conformation and, perhaps, with known good performance (either in the hunting field or competition), there is really no reason not to register any mare which is good enough to breed from, even if only at the level of mere identification (as opposed to approval or grading), as this is a start on the road to full-blown registration for her descendants in the future.

This also applies to animals of uncertain breeding but of obvious show type in categories such as hacks, hunters, working hunters, riding horses and young performance horses. If you send mares of these categories to fully registered, performance-proven stallions, their progeny can be registered (in various categories and at various levels) and their stock will command much higher prices.

So if you want to sell, and sell well, your animals must have papers. Youngstock also need to come from performance-proven parents; at least, one parent must have been a good performer. Some buyers will still buy unregistered stock on the basis of conformation, apparent temperament and, if available, pedigree, depending on its ultimate role in life, but for better prices, registration, inspection and grading are needed.

Good horses have been bred for centuries but the phenomenon of keeping accurate pedigrees is only two or three centuries old. Most breed, and even type, stud books and administering bodies were founded less than 300 years ago, and many only came into existence within living memory. The formal grading and registration now taken for granted in the warmblood world is a development of the twentieth century – and the very late twentieth century as far as the UK and Ireland are concerned.

Despite cries that it removes breeders'

autonomy to breed what they like, that it is a licence to print money and other complaints, it cannot be denied that, like it or not, papers sell horses. Documented proof of your animal's acceptance by its registry experts, its pedigree and, where applicable, its performance record, can boost its selling price to much higher levels than otherwise. The trouble and expense involved are felt by most breeders to be well worth the trouble for those who wish to sell commercially.

Fashion also plays a part in breeding to sell. Some lines, or even individual animals, will be fashionable (usually because they are good, but not always) and having such an animal's name close up in your youngster's pedigree will boost its appeal. However, there will come a point when the market, on some nebulous whim, will swing away from that particular name, so too many crosses of it will act against you. This is another good reason for line-breeding as opposed to inbreeding. Outcrossing (mating two unrelated animals) is difficult within breeds as opposed to types, as nearly all the individuals will be related, at least far back, but it will help dilute any name or family which has become unfashionable. The skill lies in recognising which names or families are going to endure and which are mere fads. You can only do this by keeping your eye on the stock and your ear to the ground.

Future trends

If the current urgings of the cognoscenti in the breeding world are heeded, we shall end up with far fewer but higher quality stallions and a much stronger broodmare band, even at grass-roots level. This can only benefit breeding. It is true that many stallions which are serving are simply not good enough, usually at studs which are trying to make what money they can out of any of their animals, and cheaply standing stallions which ought to be geldings. As registration becomes more and more the norm, such animals will in future simply not pass inspection or grading and will be unable to be used at stud for any worthwhile purpose.

The emphasis should certainly be on improving the general, all-round quality of broodmares in any breed, type or discipline. Specialist markets will dominate the breeding scene more and more, which is why it is essential to understand and plan clearly what type of foal you want to produce and what you are likely to

95

get from any mare or mares you currently have, and aim for that. Haphazard breeding is now no good unless you are going to keep and be responsible for all the animals you produce.

Overproduction

At the time of writing there are far too many horses and ponies being bred, or permitted to come into existence, as evidenced by the full pens at the markets from which most animals go into the meat and pet-food trade, and this is likely to continue to be the case for some years to come. Feral ponies of supposed mountain and moorland breeds, or at least bred there, fill these markets and few such animals seem to end up in good homes which want them and will give them fulfilling lives. They fetch only a few pounds each, and many are of nondescript conformation and certainly not of good, pure-bred type of any breed of native, and it would surely be better if they had never seen the light of day.

Small 'hobby' breeders who breed indiscriminately just for the fun of having a little foal frisking about their paddocks overlook the fact that little foals grow into big, boisterous, hungry ponies and horses and, if they are not skilfully trained, they can become dangerous and useless. They, too, are on the road to the meat factory – it just takes them a little longer. Small can be beautiful when it comes to studs and horse breeding, but you should always know what you are aiming for, and to aim for a known market with high-class stock – with papers – if you want to get a reasonable price for your stock and ensure a useful life for it.

Probably one of the worst things a breeder who is having problems selling stock can do is to retain it and breed further from it in the hope of selling in the future, yet so many do this. It should be obvious that this is compounding a serious problem. A surplus of animals reduces their selling price and those which end up in the markets, as most unwanted stock does, are usually unkempt and in poor condition by that time, which does nothing for the horse world's image in general and discredits the carefully laid plans of the registries and breed societies.

Breeding unregisterable animals, or allowing feral animals to reproduce unchecked, floods the market every year with scrub 'mongrel' animals of which only a very few will find good homes. Most end up in tins, some are bought by well-meaning individuals and some are taken by equine charities, possibly with a view to rehoming. However, if they are not good specimens, it will be hard to rehome them except as companions (and there are already too many of them) because they cannot stand up to work.

Time and again, we come back to the same unavoidable answer to these problems: *Breed far fewer, much better animals across the board.* If we only had half as many horses and ponies in the country as we have now, and if they were only 50 per cent better in quality than they are, the situation would be improved tremendously.

Breeders are responsible for lives, and animals in this world come a poor second to human beings. Our horses and ponies live as slaves in that we are ultimately responsible for their fates, their lives and their deaths. It is unrealistic to expect breeders to accept responsibility for their produce for the whole of their lives, but a letter in an equestrian magazine some time ago suggested that conscientious breeders should not sell their animals but lease them so that they could ensure that they always had good homes. In the long run, the writer calculated quite convincingly, breeders would receive far more income from each animal than by selling it, their consciences would be clear, they would have control over the animal's final years or days and it would also encourage them to breed fewer, better horses, as they would be easier to place and would command more 'rent'. Riders would also not have the relatively large outlay of buying a horse and could return it without the hassle of selling it should they become unable or unwilling to keep it.

It is true that the fact of a horse's being potentially top-class does not guarantee that it is going to be treated humanely all its life. Often, the most talented are those that are bled dry till they can give no more and are then abused in an effort to squeeze even more out of them. Horses and ponies capable of winning in any sphere can be subjected to appalling abuse even in disciplines which do not, at first sight, seem to be particularly demanding such as dressage and showing. Of course, abuse is not the prerogative of the competition world and, idealistic though it may seem, a leasing scheme by concerned breeders seems well worth considering.

However, in the real world, perhaps the best we can hope for is that conscientious breeders will breed far fewer and far better animals and try to ensure that the people they sell them to are worthy of them. This, at least, would be a big step in the right direction.

Chapter 12

Financial Matters

Even taking into account council tax and other rates on landed properties, there is no doubt that the cheapest way to keep a horse or to breed horses is to own your own land and ancillary facilities and to do the work yourself. But whilst many small or novice breeders may be doing the work themselves, not employing stud staff at this stage, they may not all have their own land, stables and so on.

If you do have your own place, and especially if you have previously kept horses at livery, even DIY livery, you will know that you can easily keep two or even three horses at home for the price of one in rented accommodation or on paid livery. You have the advantage of being able to arrange your premises more or less as you wish (given planning permission and money) in order to create facilities or make the best use of what is there. You can also look after your horses exactly as you see fit, turn them out when you want, bring them in when you want, feed and bed them on products of your own choice and establish your own routine, as well as having the peace of mind of knowing that no one is possibly mistreating them in your absence or failing to carry out your careful and simple instructions. Keeping horses on other people's property can be a singularly unrewarding, expensive and frustrating experience. It can be extremely risky, too, to make yourself responsible for the care and wellbeing of even one horse, one mare and foal or more, and suddenly to have your horses made homeless through no fault of your own and for various reasons beyond your control.

Before starting to breed horses, therefore, probably the most important things to consider (apart from being sure you are actually knowledgeable enough about horses or ponies in general) are:

- where you are going to keep them
- for how long this accommodation will be available
- how much it is going to cost.

There are, of course, other considerations, but without accommodation, facilities and money, you cannot even begin to make your dream come true. If you go ahead anyway, trusting to luck too much and blithely deciding to cross any bridges when you come to them, you could find that your dream becomes a nightmare in no time at all.

Accommodation and facilities

You will certainly need productive, clean, well-drained and well-sheltered pastureland for your horses or ponies, even just for one mare or foal, because, as I have said, youngsters do not develop properly unless they have the facility to be on the move for most of the time. Some people keep horses 'up and in' for six months of the year, from late autumn to early or mid-spring, because their land is too wet for the horses to go out for long if at all, and they have failed to rectify the drainage or to create other adequate exercise facilities for them. Such confinement is not the way to produce well-grown, strong, mentally and physically healthy youngstock. If you make sure your land is suitable for livestock in the first place – neither heavy clay which waterlogs in winter and bakes hard in summer, nor very free-draining land which becomes a dust-bowl in summer – there will be no need to spend large amounts of money on installing underground drainage systems or, conversely, irrigation systems. It would probably be cheaper in the long run, anyway, to sell such land and buy something better.

Any type of horse or pony which is sensitive to weather (usually those with a good deal of Thoroughbred or Arab in them) will probably not be keen to spend long in a cold, windswept field in winter and these conditions are certainly bad for young foals. What is needed at any time of year is well-drained, productive land which will still retain enough moisture in very dry weather to maintain grass growth, and which is also well sheltered.

Man-made field shelters or the facility for the horses to go into and out of a barn or some other

97

The inside of an American barn stabling system on a Thoroughbred stud. Some studs are reverting to conventional outdoor looseboxes with a staff corridor behind as it is felt that they provide a healthier environment. However, existing facilities can have extra ventilation outlets fitted such as ridge-roof ventilators and louvres high up on the walls, not to mention opening doors or windows in the back walls of the boxes. These not only help with the airflow but also give horses another important viewpoint in an otherwise psychologically sterile environment. Horses are intelligent animals and need mental stimulation as well as physical care.

building (even their own stables depending on the lie of the premises) are excellent, but good natural shelter on the windward side of the field – thick, high hedges and shelter belts of trees – is also a boon to any horse pasture, particularly that which is meant for breeding stock. Horses will use shelter just as much in summer, to get away from sun and flies, as they will in winter to escape from driving rain and relentless wind.

As for the amount of land, the general rule of thumb is that, for year-round use and to enable you to rest and rotate your grazing for correct management, you need a minimum of 1 hectare (2 acres) of good land for the first horse and half as much again for each additional horse or foal.

Bearing in mind that foals need space to exercise, think and develop, and that there may be times when the weather or the land are really too bad to turn them out (perhaps on small, hard-used areas in winter and during abnormally wet periods), you will need to create alternative exercise space for young animals because one of the worst things you can do to them is to confine them to loose boxes for twenty

or so hours a day and believe that a half-hour run on an outdoor manège once or twice a day (which is all some livery-yard owners may permit) is enough to enable them to let off steam. It might let them do that, but it will not help create the healthy stress and development youngsters need if they are to mature into well-proportioned horses of strong, healthy constitution.

An invaluable facility, therefore, is a well-sheltered but well-ventilated area with a safe ground surface (earth, sand – but not salty beach sand in case they take to licking it – wood-chips, shavings, straw or some other cushioning, non-slip, safe surface), perhaps about the size of a good outdoor school so that the foal can get up a canter and kick around. This can be indoors, outdoors or half and half, just so long as there is room to move freely and somewhere to get away from the weather. And any companions turned out together must obviously be compatible.

Safe fencing is also needed and, for breeding stock, the safest is probably thick, high, non-poisonous hedging, thick all the way to the ground with no possibility whatsoever of animals, including small, inquisitive foals, pushing through it. Five-rail wooden post-and-rail fencing is also good provided the bottom rail is about 30cm (1ft) from the ground. This will be high enough to stop a rolling foal ending up on the wrong side of the fence and the gaps between the rails will be too small to allow it to scramble through on an exploratory foray of the surrounding area. The top rail should be at least the height of an adult animal's withers.

Concrete posts and wire, and especially barbed wire, should be avoided like the plague for breeding stock, as should sheep netting. Proprietory types of fencing, such as very small diamond mesh are now available, and can be very good. Rubber and plastic flexible rail fencing may be attractive to foals for teething and investigative purposes, and electric fencing should, in my view, never be used for youngstock. The same safety rules apply to gates and, if you are forced to use them, slip-rails. (You will need five slip-rails to stop the foal getting through, not the usual two.) Gates with cross-members can be dangerous as they easily trap

small hooves. They should be filled in on the field side with a strong metal grille, not wire mesh or netting which is too weak and flexible.

All hedging and fencing should be on the field side of any ditches or dykes so that there is no chance of horses, especially youngsters, falling into them and possibly dying from stress, drowning or pneumonia as a result of pressure on their lungs.

A common complaint about turning out horses for exercise on non-grassed areas is that it is a waste of time because they just stand there moping. They might indeed do so after the first few minutes if there is no food for them to eat. The idea is to make sure they have ample roughage (hay, haylage or forage feed) in safe, fixed containers (not hay-nets for youngstock and nothing with sharp corners or spaces in which they could catch their feet) and also water, so that they can eat when they want, then go off and play, and come back for more food. Although no one would pretend that this is as good as pasture from their point of view, it is far better than confinement in a loose box or exposure on a mud patch or baked hard dust-

---

Indoor looseboxes may not be as well ventilated as formerly believed. In some poorly-ventilated facilities, the stale air pools inside the boxes but a lower door of strong metal mesh, like this, helps overcome, the problem and also helps to lessen the closed-in feeling experienced by the occupants. It is likely that many more horses than we realise are claustrophobic due to their natural evolution.

An example of ideal, traditional post and rail fencing for breeding stock. There are four rails which are ideal for horse foals, the bottom one is low enough to prevent even a young foal rolling underneath and ending up on the other side in a panic, the top rail is the height of the mares' backs and is flush with the tops of the posts, which are sloped back to permit rain run-off (which will prolong the life of the wood) and to remove any projections which could seriously injure an animal attempting to jump out. In addition, the rails are on the correct side of the posts, inside the field, which prevents shoulder impact injuries from exposed posts to animals galloping along the fence line. (Peter Sweet)

bowl, however big. Very hard ground (frozen or baked) is bad for youngsters' feet and legs if they gallop around on it, and hardened ruts and pocks due to land having been poached and not subsequently rolled is also dangerous.

Most establishments used for horses and ponies will have some area where a non-grass turnout facility could be created with a bit of imagination and moderate expense. You have a lot of leeway for using your initiative, provided you bear in mind the horses' comfort, safety and enjoyment and the youngsters' need to move and learn to use their bodies. Standing around for most of their time in forced confinement is one of the worst things you can inflict on youngstock. Several hours a day on a decent play area will do when the available grassland is not suitable, perhaps just bringing them in at night in winter or during the day in a hot summer.

Only you can assess the suitability of your facilities. Native ponies and cobs may well be better off living out all year round given good land and shelter. Animals with any 'blood' in them will almost certainly need to be brought in at night in winter, if only into a bedded barn, but given good land and shelter they should be able

to stay out for the rest of the year, which would be better for them.

It is also perfectly possible to keep horses without stables, even though the first thing most owners think they are going to need is a stable! Actually, the most important thing is exercise facilities and land, and horses do very well – and are usually happier – in communal yards, provided they are compatible. Communal yards are also much easier to manage, involving a lot less work than individual stables.

If your mare is foaling at home, however, and is not a native type which will foal outdoors, you will probably feel safer with a good-sized foaling box for her. The advised minimum measurement for a foaling box used to be a little over 4m (14ft) square. Nowadays, modern research shows that horses prefer much bigger stables (around 4.5m/15ft square for one horse); if you can manage anything bigger than that old guideline by all means do so. Obviously, a mare and foal together (which they will be for at least six months) cannot safely use a conventional stable, so if you only have one mare or maybe two you may be able to let them keep their foaling boxes for permanent use together. On large studs, it is

Wide, sweeping lanes between paddocks are a great asset when moving horses or maintenance vehicles around the stud if you have the space . . .

. . . but on a smaller establishment narrower lanes suffice. A slip-gate at the ends of lanes is a safety feature should led animals break loose. This one has been left down for the photograph to show how it slots into the holders on the posts.

normal for a mare and foal to have the foaling box for a few days, then be moved to one that is larger than normal but smaller than the foaling box. As a 'small' breeder, you may not need to do this.

In addition to these requirements, you will need all the other general horse facilities such as a reliable, ample, clean water supply (probably a piped mains supply for safety and reliable purity with ponds, dykes and streams safely fenced off), storage for tack, feed, bedding, hay or haylage, muck heap or disposal facilities, and so on.

It is easy to see that the specialised facilities needed for breeding stock are unlikely to be found on non-specialised stable premises such as ordinary livery yards, rented facilities or ordinary livestock (non-horse) farms. I have known a great many heart-breaking accidents and fatalities occur simply because people were keeping their animals on other people's premises and were not allowed to change them to make them suitable for breeding stock. And even if the facilities are suitable, it is usual for boarders to have their use of them severely restricted. An extremely common and serious deficiency with other people's premises is the lack of suitable grazing, or simply a refusal to let you use it.

It may sound hard, but unless you are boarding your animals on a reputable stud or other suitable place where you are sure of safe facilities, ample grazing and turnout, and expert, conscientious and co-operative care, you really

Conventional stabling is by no means essential. These pony mares are happier in this open-fronted building where they can enjoy each other's company.

need your own place, or rented or leased premises which you can alter and use as needed, before you can say you have suitable facilities for breeding horses or ponies. The game is risky enough as it is without stacking the odds against yourself.

Will you want to expand?

Many people who go into breeding are severely bitten by the bug and want to expand. Apart from the responsibility involved in bringing into the world more and more lives (including more and more mouths to feed, more and more vaccinations and worming doses and more and more feet to trim), you need to reconsider the cost, the facilities and the market for your stock. You cannot keep ten horses on a place meant for two so it is important to have your accommodation and finances, and possibly suitable staff, organised before you start buying extra mares or putting your fillies in foal.

As I have said, the market is flooded with excellent horses and ponies for sale, many of which can barely be given away, so a talent for objectively predicting future markets and the

public's requirement for the type of stock you are breeding is invaluable. Remember that followers (youngstock not ready to sell profitably, usually because they are too young to have been backed and ridden) mount up alarmingly. These days, most people expect to be able to buy 'instant rides' which they will not get until an animal is about four years old. Do you have the strength, skill, money and time to cope with boisterous youngsters, particularly entire colts, until they are old enough to back?

Although they can be backed and ridden away (ridden lightly) at three years old, they cannot do any worthwhile work at this age but many buyers these days do not understand or accept this fact. This is why there are so many wrecked youngsters around. All this means that, unless you breed exceptionally high quality, registered youngsters, and can be sure of selling to knowledgeable owners who will wait for them to mature (even if you sell them at three to owners capable of backing and schooling them), you will not be able to sell your animals at a profit, or even a break-even price, unless they are rideable, reasonably well schooled and safe in traffic. You cannot achieve this with a three-year-old, so you will be keeping all your produce until it is around four years old and you will need the skill yourself, or the cash to pay some such person, to back and school them all – quite a sobering thought.

Of course, you may be in breeding for the love of it and be willing to sell youngsters at a loss, simply recouping part of your expenses, or keeping them for your own and your family's use. It all depends on your outlook.

Knowledge, time and money

I have already said that any potential breeder of horses or ponies should be very knowledgeable about horses in general before embarking on this very specialised field. Animals are easily ruined for life by bad treatment and management on the home stud, sometimes the result of ignorance rather than ill-will, often because the breeders are completely taken by aback by the way foals and youngsters behave. They are, in effect, wild animals, but novice breeders often seem to expect them to come out into the world fully accepting of humans, automatically leading well

in hand, willing to pick up their feet, move over in the stable, be caught in the paddock and so on.

Nothing could be further from the truth; young foals (unless imprinted or very frequently handled from birth) naturally avoid humans, running behind their dams at the approach of people. They have no idea about wearing a foal-slip, they are instinctively frightened of surrendering their feet for attention or their heads for leading, they do not understand English and generally behave as if they are being disobedient, whereas in fact they are simply behaving like young foals. They have to be taught how we want them to behave and, although they will follow their dams' examples and probably pick up their attitudes to humans for better or for worse, they need correct handling and training from a young age if they are not to become very hard to handle, even dangerous, as they grow in size and strength – which they do rapidly.

Of course, everyone has to start somewhere, but it would be a good idea to help out at a stud under expert supervision, if you can find one where the staff and owners have the time to

This barn opening onto a paddock is a wonderful facility. Horses can come and go as they wish according to their need for shelter and the high doors can be closed to keep them in, if necessary. (Peter Sweet)

teach you, or take a practical course in stud management and the handling of youngstock before breeding your own. Books, magazine articles and correspondence courses are an excellent source of information which you can absorb in your own time and at your own pace, but nothing can take the place of practical, hands-on experience under the supervision of people who really know what they are doing.

You may feel at first that there will be no need to spend extra time with your breeding stock than before you were a breeder. However, you will probably find the physical and psychological changes in your mare or mares fascinating and you may spend a fair amount of time just watching them, which is never a waste.

Once foaling draws near, and particularly after the foal is born, you will be around your mare and foal a good deal just for your own peace of mind and curiosity. The foal will also need handling and schooling in the ways of a socialised horse, which all takes time. However, if you have time to ride, groom and exercise a riding horse, you will certainly have time for your mare and foal with no ridden exercise needed and no time-consuming strapping or full grooming. The extra time you give will consist largely of satisfying your own fascination and anxieties rather than entailing a lot of extra work.

Your expenditure will start increasing as soon as you decide to put your mare in foal because she will need to have her vaccinations and worming programmes brought up to date if they are not already and will need to have various swab tests done before going to stud. The cost of the vet's visits to do these various jobs is not insignificant and you have postage, laboratory fees and possibly VAT on top. If there are any problems, more expense will be involved in putting things right.

You may have to spend money on getting your premises into shape for breeding stock, and you will also be spending a good deal of money and time travelling round looking at stallions to choose the right mate. Once the mare is at stud (remember the cost of getting her there), there may be VAT on the stallion's covering fee and, if it is a big place, maybe on livery services. There will be tests and scans at the stud and possibly other veterinary attention.

Check also the exact arrangements for paying the covering fee. Some studs work a 'no foal, no fee' scheme, others charge no fee if your mare is barren on 1 October, and yet others a 'no foal free return' (to the stallion) scheme. You may have to pay half of the fee up front and the other half if the mare is tested in foal by 1 October. There are all sorts of arrangements, so make sure you can

A catching pen at the entrance to paddocks is a big help whether bringing in or turning out. The outer gate has been left open for the photograph: note the safe, solid side-boarding. Animals are led through one gate which is closed behind them before the next one is opened. This helps prevent barging or unruliness and, of course, helps stop horses escaping. (Peter Sweet)

This catching pen is fitted with safe and strong metal mesh gates which completely prevent an animal getting a leg through the bars. The height of the gates has been increased, to discourage jumping out, by means of grille extensions painted white so that the horses are in no doubt that they are there. Gateways are a frequent source of trouble with horses, who gather there at any time when they are expecting to be fed or brought in.

meet these financial obligations before commiting yourself.

Once the foal arrives, there will be extra costs in keeping both the dam and the foal – extra worming, extra farriery, extra vaccinations and possibly veterinary checks, plus extra tack for the foal and, once it begins a more independent life, extra feed and bedding. The costs will not go down! The final insult is usually a rise in your insurance premium.

If you are foaling your mare at home but want your vet present or on call, this will cost you, too, so check how much it is likely to be. You will need to cover registration costs for the foal and maybe the mare (bearing in mind that your stock will be worth more if it is registered itself and from registered parents) and, looking on the black side, if there are problems there will surely be yet more veterinary fees plus the cost involved in hand-rearing a foal (your own or an orphan with your mare), or the cost of transporting one or the other to its new partner in life.

So as you can see, setting up in a breeding operation, even with only one mare, is an expensive undertaking, not a quick money-maker. Many a would-be breeder has been put off after wisely doing the sums before the event,

which is all to the good.

Veterinary surgeons

Whether or not you are an experienced breeder, your vet will be your indispensable helper. Very few people go through their horse-owning lives never needing a vet, even if they never vaccinate their horses, and it is important to get the right person. You need someone you can get along with and talk to. Vets in general are much more communicative than they used to be years ago; at one time, you took their word for things, did as you were told, did not ask questions, paid their bills and that was that. Very often, you did not even get a clear explanation of what was wrong with your animal, what treatment the vet was giving, what drugs were being used or what the outcome was likely to be! Now, professional people are no longer on pedestals; they are regarded as just human beings, but with specialised knowledge and skills for which you, the client or customer, are paying. Vets are therefore far more forthcoming today and are willing to spend time explaining what is wrong, what treatment they are giving and why, what you must do to support the treatment and what the prognosis (outlook) is likely to be. If there are

105

options, you are much more likely to be brought into discussions than in the past so that a joint decision about treatment can be made.

It is important to ensure that you have an insurance policy which at least covers veterinary fees for your animals, especially where something as potentially risky as breeding is concerned. Remember, however, that insurance companies are like bookmakers – they only like betting on dead certs. The riskier your situation appears to them and the higher the value of your animals, the higher the premium and, in some cases, you may not get cover at all or find it so restricted that it is not worth paying for.

It also has to be said that not all insurance companies appear to treat their clients ethically. Read small print meticulously and get written explanations (which you understand) of any points which are not clear. A personal recommendation of a good company is better than the hype you read in advertisements. If you find a particular company's brochures in your vet's surgery or office, it is likely that he or she has found them good to deal with.

All vets have the same basic training in the treatment of all types of animal. The training is long and difficult and can be arduous and dangerous, as can the work. There is no particular qualification needed to work as a specialist equine vet; once qualified, vets can specialise in one particular type of practice, or one species of animal, if they wish. But those who belong to the British Equine Veterinary Association in the UK, for example, or who have a large animal practice or, more particularly, an equine practice, can be expected to have a special interest in and experience of horses and ponies, so it seems a good idea to use such a practice. Local knowledge, experience and recommendation will also help you choose. Horse owners are always happy to talk about their vets!

Farriers

Most horse owners see far more of their farriers than they do of their vets Your farrier will be coming about every six weeks to trim your mare's feet and, in short time, your foal's and again it is essential that you have someone you can talk to and who will discuss matters with you rather than just adopting an 'I know best, it's nothing to do with you' attitude. Really competent farriers who are also professional in their treatment of their clients, horses and work are still too thin on the ground, at least in the UK,

but they are essential when it comes to ensuring good limb conformation, action and soundness of youngstock. Also, broodmares' feet come under a great deal of stress and strain and need to be kept well balanced and trimmed.

If your area really is poorly supplied with good farriers, it is important that you make enquiries in the surrounding district to find someone suitable who will come and visit your yard regularly to ensure the essential correct care of your animals' feet and legs.

Nutritionists

Although there are at present no truly independent nutritionists working in the UK, it is a good idea to build up a good working relationship with the nutritionist working for the company whose feeds you use. Not many people these days use no branded feeds at all, but those who do can get relevant dietary information from their vet.

Reputable feed companies (not always the largest) will have at least one nutritionist available at certain times to answer general queries on nutrition. As I have said, it is not good commercial practice these days for nutritionists to be seen to be pushing their company's feeds and products to the exclusion of everyone else's. Few companies make absolutely everything you may enquire about – for example, some make and market feeds, some only deal in supplements – and you should be able to get unbiased information about nutrition in general, about that company's products and what they recommend from elsewhere if they cannot supply it. Of course, they want your business, but they know they are more likely to get it if they operate an open-minded policy to ensure that you get the best products and advice for your circumstances.

Possibly your best plan is to use good, branded feeds specifically formulated for particular categories of breeding stock (pregnant mares, lactating mares, foals, yearlings and so on) and leave the matter of supplements to your vet, who you know is unbiased. If you have your soil and herbage and your roughage (hay or haylage) analysed, or use a forage feed with the analysis on the bag, and top up if needed with 'short' feeds or concentrates from a good company, you should be able to get all the information and advice you need from your vet as to whether or not a supplement is needed to balance your horses' existing diets.

Chapter 13

Your Foundation Stock

The veterinary aspects of the good and bad qualities of breeding stock were dealt with in Part 1, so in this chapter I shall be discussing more practical matters relating to both mares and stallions.

Soundness

As a novice breeder, it is safest for you to breed only from sound stock or from stock which you know is unsound because of a genuine accident. Any animal that is unsound for other health reasons should normally be avoided unless a veterinary surgeon expresses the opinion that the condition is unlikely to affect breeding performance and is not heritable (able to be passed on). Many performance animals retire to stud with leg problems, and there is something to be said for the argument that unreasonable stress such as is often placed on horses during competition is not the horse's fault and even that susceptibility to leg problems can be due to poor management such as inadequate fitness, inconsiderate work and so on (as can allergic conditions such as COPD or Chronic Obstructive Pulmonary Disease, commonly known as broken wind). This, however, is a grey area and is best left to experts to cope with.

Breeding from unsound animals is tempting for relatively short-term gain, but it will do nothing for the reputation of your stock when it becomes known that they usually develop health or soundness problems or cannot stand up to the rigours of normal training. If you are breeding for yourself, you will be in for a bitter disappointment when your own home-bred youngster proves to be unable to stand the pace – unless that pace is just casual hacking and light pleasure riding.

Conformation and action

The topic of conformation is a huge one, and I have covered it in some detail in my book *Conformation for the Purpose*. Basically, you should aim to use animals of as good conformation as you can possibly find, bearing in mind that none is perfect. No animal, male or female, with any serious or significant conformation fault should ever be bred from in the hope that the fault will be cancelled out by a mate which is strong in that department. Some faults can certainly be bred out, as can some diseases and other weaknesses, but this is a job for an experienced breeder, one which will involve meticulous attention to bloodlines, outcrossing, inbreeding and line-breeding. As the science of genetic engineering develops, other means of removing faults from breeding stock may become available to the horse world but, for most breeders, it is far safer, cheaper and less heart-breaking to avoid stock with faults in the first place.

Crooked action is something else to avoid. Equally strict standards should be applied to an animal's action as to its conformation (horses were not born to stand still). Animals can be born and develop to maturity with good, straight action only to develop crooked action in one or more legs under the stress of work: for instance, an injury may leave a tendon permanently weakened, which causes the movement of the leg to change, or concussion can cause bony deposits around a joint which affect its action. This may show a weakness in the horse's physique, although poor farriery can also cause it (including the owner not calling in the farrier often enough), as can overwork, working when 'unlevel' or 'not right' (euphemisms for lame), which overstresses the part, and injuries.

Animals which are in weak or thin condition, or which have been incorrectly worked, resulting in the development of the wrong muscles for athletic work, may also appear to have poor conformation and action, but once they are in better condition and have had several months of competent schooling leading to correct muscle development, they often improve greatly and show their true colours.

A good way of knowing whether or not a stallion throws offspring with his own qualities

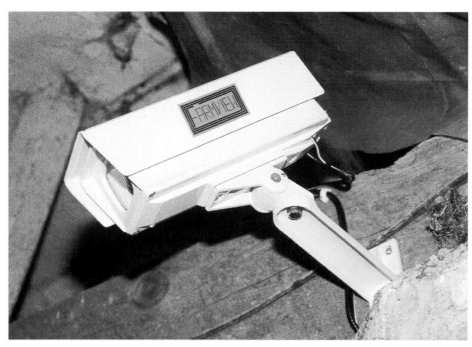

A closed circuit television camera installed for observing a mare without disturbing her. This facility is now found on even small studs. (Vanessa Britton)

from a mixed bag of mares is to go and see as many of them as you can. When you reach the point when you can pick out his stock among a bunch of others, you will know you are considering a pre-potent stallion who consistently produces foals in his image. This also applies to mares, of course, and if you are buying a broodmare rather than using a mare you already have, look at her offspring, too. These will be far less numerous, of course, but many people still place all the emphasis on the stallion while the mare is just as important in passing on genes and traits. An interesting old saying says that the biggest compliment you can pay a stallion is to say that he throws even better stock than himself, no matter what sort of mare he is put to; if you can find one of these you should be onto a winner.

Performance history

You stand more chance of breeding saleable, talented youngstock with good athletic potential, able to command a fair price, if you make sure that both their parents are proven good performers *and producers of performers*. More and more performance stallions of all breeds are becoming available now and studs are proud to show their records. As for your mare, 'performance' does not necessarily mean Olympic or world-level winnings or even actual competition; if she has been a successful hunter

in galloping, jumping country this is very much in her favour if you are breeding competition horses or ponies. If she has excelled in the showring or at Riding Club or Pony Club events, been consistently in the ribbons at local and especially county show level in the appropriate classes (breed, type, discipline); or even just a reliable, safe, active hack over varied terrain involving ditches, hazards, streams and hedges, which will take anybody anywhere (a pearl without price to some people); she is a proven performer; and the use of a high-level performance-proven stallion should more or less ensure that her stock will perform better than she did.

Performance ability in your mare's close relatives is also an advantage, particularly if she has never excelled herself. As the stallion is concerned, close relatives (parents and siblings) with performance ability, and especially offspring which consistently do well and which have produced good performers themselves, show that dominant genes for the talents you are looking for exist in his family. You need to make sure that he is not a freak high-class performer with no others in his bloodline (they do occur) as you could gain no performance genes in your own stock.

It is cheaper to use a stallion from a family known to produce good performers but who has produced none himself as yet (such as a young horse in his first few seasons), particularly if his parents and/or siblings have done well, but it is not quite such a safe bet. Young stallions which are retired to stud because of accidents, without having proved themselves but which are from performance-proven families, come into this category and it is worth considering, at least.

It should be stressed that, in this context, 'performance' means ability in the discipline in which you are interested. Whether you want to produce safe children's riding ponies, in-hand showing stock, ridden show animals, Riding Club mounts, Pony Club all-rounders, effective hunters in the field, show jumpers, eventers, dressage horses, show or competition driving horses, racehorses, endurance horses, pleasure horses and hacks – the list nowadays seems endless – as a breeder you naturally want to see performance history. This can be on a database with an official administrative body or in careful hand-kept records of owners or breeders of your animals' ancestors. You can manage without all this but not if you want to breed to sell and eventually not only break even but also make a profit.

Health history

As far as your own mare is concerned, you will have all the knowledge you need about her health and constitution, but you may find it

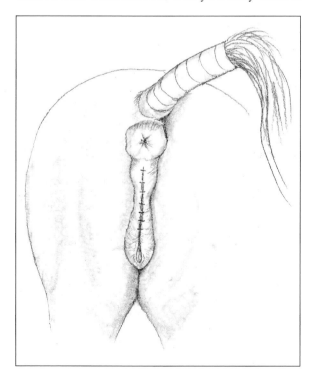

extremely difficult to get hold of this kind of information concerning a mare you are buying for breeding or a stallion you are considering using. Probably the only way you will get near the truth is to ask around among people who know the animals concerned, if you do not know them yourself. You should in any case have any potential purchases vetted before committing yourself and have a vet check over your own animals for breeding suitability (including the passing on of faults and diseases).

From the point of view of a mare you have owned for some time, if you have found that, despite good management and veterinary care, she frequently falls ill or simply never really thrives, perhaps you should seriously consider whether she should be bred from.

Temperament

Some of the most meticulous and calculating breeders in the world, the continental European warmblood breeders, have for generations made good temperament a major priority in their stock, along with conformation, action and performance. The result is that most warmbloods have good natures and are very trainable, although there are exceptions. Breeders of other breeds and types worldwide do not seem to have placed quite so much emphasis specifically on temperament with the probable exception of American Saddle Horse breeders, whose breed is renowned for its friendly, tolerant temperament. Many breeds are noted for good temperament but this is often a by-product of the policy of discarding bad-tempered individuals rather than of actively selecting kind ones.

One breed whose temperament does not come into the breeding equation at all is the Thoroughbred when it is bred for racing. Indeed, by far the most important quality considered when buying yearlings for racing appears to be

The vulva of a mare which has had a Caslick's operation. Mares whose anuses are sunken back behind the line of the vulva contaminate the vulval lips with droppings which can lead to infection spreading up the vagina and into the uterus. Sometimes, after having had many foals, or due to old age or injury to the area during foaling, the vulval lips may become too loose to form the normal seal of closed vulval lips. In these cases, the operation can be performed to improve the seal: the vet, will inject a local anaesthetic and trim the outer skin from the lips, stitch them together to a point just below the pelvic rim and subsequently remove the stitches after the area has healed. This, of course, does not prevent the mare's staling but the operation must be reversed if the mare is to be used as a broodmare again.

speed in the family. The more astute purchasers pay considerable attention to conformation but temperament does not rate a thought; the professionals who will be doing the caring and training are expected to be able to cope with a difficult temperament. Because of this, many Thoroughbreds inherit highly strung, opinionated, sensitive, moody and occasionally vicious temperaments which even 'professional amateurs' do not want to have to deal with. Along with that, however, often go the 'heart', courage, determination and guts under pressure so often lacking in other breeds. This is why Thoroughbreds more than any other breed today are used for upgrading and adding quality to other breeds and types. Arabs, with their intelligence, courage and normally good natures, are also used to add panache and flair, if to a lesser extent nowadays, and they are one of the ancestors of the Thoroughbred, of course.

Unlike their counterparts in racing, those buying Thoroughbreds for breeding competition horses do rate temperament very highly as more and more of them are absorbed into the various warmblood breeds and sporthorse types. In that world, only horses with a good temperament, constitution, conformation and action will be considered and it should be stressed that there are many Thoroughbreds with perfectly good temperaments.

There were temperament problems with the over-use of indiscriminately selected small Thoroughbreds some years ago, in the breeding of the British Riding Pony. Despite the fact that the aim was a children's riding pony, many of the children riding the resultant showring miniature Thoroughbreds were themselves 'professional' show riders able to cope with a pony of dubious temperament. Nowadays, after much justifiable ill-feeling against this type of pony in particular because of the poor temperament and loss of hardiness, the ponies we see in the showring have much more equable temperaments and are more suitable for children in general. After the

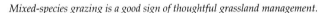

Mixed-species grazing is a good sign of thoughtful grassland management.

An indoor barn, with Yorkshire boarding for excellent ventilation without draughts, is a valuable facility, particularly for mares with foals born early in the year. It is also a welcome haven from sun and flies in high summer. The doors at the far end of this barn can be left open and lead directly onto a grass paddock.

reintroduction of more native pony blood, they are also more pony-like if still obviously with a 'blood' background. It takes a skilled breeder to produce a quality children's show pony which is not just a small Thoroughbred and which has a temperament safe for children. One now often has to go back several generations in the pedigrees of show Riding Ponies before coming across a pure Thoroughbred or Arab.

Arabian stallions have been used in the breeding of children's ponies and generally have more equable temperaments than Thoroughbreds unless they are badly handled. Arabs as a breed are renowned for their kind, intelligent natures but they will not tolerate or work for idiots and should by no means be thought of as a poor man's Thoroughbred. Unfortunately, today's showring fashion in the Arab world is to encourage animals of a conformation which must be making the earlier breeders turn in their graves, as must the way in which they are shown, with heads in the air, hind

legs outstretched behind and spines consequently over-stressed at the very spot where weight would be carried under saddle. The fact that many of these in-hand show animals never carry weight is irrelevant.

Arabs are now used extensively for endurance riding and Arab racing, which is attracting more and more professionals, often from Thoroughbred racing. An eye will need to be kept on horses from the latter discipline in case, in generations to come, their temperaments deteriorate as a result of breeding only for speed, as in the Thoroughbred.

It seems that specialisation for showing is already pushing the traditional good temperament of the Arab out of consideration in some cases, so novice breeders should satisfy themselves as to the temperament of any Arabians and their crosses which they are considering using or buying for breeding. Pay a personal visit to the stud or owner and, in addition to your normal inspection of conformation and action, pay particular attention to the animal's apparent temperament. If the owner or staff will not let you handle the animals, be suspicious – there must be a good reason!

This advice can obviously apply to any breed or type of horse or pony. You should never take for granted that an animal has a good temperament, even if you are told it has. A majority of horse owners are amateurs who have horses and ponies for fun, if also, in many cases, for the challenge and achievement of competing. They often have no professional training and are easily put off the sport by animals with unco-operative or even dangerous temperaments. As this sort of owner will probably form the biggest part of your market (or the market of the person to whom you sell your green youngsters), the temperament of your stock is extremely important.

Apart from the fact that you yourself will enjoy breeding more if you breed animals you can be friends with, if you gain a reputation for producing animals with good temperaments and which are friendly, co-operative, good to handle, quick to learn and a pleasure to be with, you should find a reasonable market for them. Not many 'ordinary' owners, or those who bring on youngsters for sale to such people, want awkward, unco-operative or downright nasty animals. The 'middle men' may well be able to cope with them but poor temperament can be hard to disguise and they, too, want to maintain a reputation for selling on good stock.

By placing good temperament as high on your list of priorities as health, conformation, action, soundness and performance, you will be doing the horse world a big favour and ensuring a good reputation for yourself and, it is hoped, a ready market for your stock.

Pedigree

The science of pedigree analysis is one you either have a bent for or not. Counting up how many crosses there are of a particular animal and calculating how much of its blood there will be in your foal or foals can be an absorbing winter's evening pastime for those with a statistical bent.

Generally speaking, however, it is said that any animal more than three generations back in a pedigree will have so little influence on your foal that it is not worth bothering about. Having said that, I once saw an old Thoroughbred stallion, one of whose ancestors several generations back

(amounting to 100 years or so) was the famous racehorse, St Simon. As I said earlier, I have many old books containing pictures of St Simon and am very familiar with his appearance, and when this descendant of his walked into the arena where he was to be used in a demonstration, I could have sworn it was St Simon!

Although I am not an authority on present-day Arabs, I have noticed that the descendants of the Crabbet-bred stallion Rifari usually look like him and there must be many other examples in this and other breeds. It is an interesting pointer to look for. Because genes have been passed down for millions of years, they must have some influence, particularly if a particular cross is repeated twice or more.

For practical purposes, the main things to look for in the pedigree of your mare and proposed stallion, are animals of proven performance ability and those which you have particularly admired. The number of times an animal (male or female) appears in your planned foal's pedigree should also be considered, particularly if those appearances (crosses) are within three generations. For safety's sake, it is probably a good idea, until you gain more experience of practical genetics and relationships, to avoid any repetitions within at least three generations. With some breeds this can be tricky, but there are enough good animals out there for you to be able to avoid recent crosses until you are more certain and confident of their effects.

Age

Breeders have for long speculated about whether the age of a mare at conception and foaling makes any difference to the quality of her stock. It is generally felt that old mares do not produce their best stock and nor do very young ones, but it seems that a mare's breeding career as far as producing high-class stock is concerned may be shorter than we think.

An interesting piece of work done by Jane Barron of the Warwickshire College of Agriculture and published in the British publication, *The Equine Veterinary Journal*, seems to indicate that mares produce their best offspring during a short period of prime production years (she studied Thoroughbreds but there is no reason why her findings should not apply to other breeds). She checked 100 mares who had each had at least eight live foals and compared their Timeform ratings when they raced. (Timeform ratings are an industry rating of an animal's racing ability.) She found that the rating given to foals from six-year-old mares was 99, gradually increasing to 106 for the foals out of nine-year-old mares and then decreasing to 84 for foals out of eighteen-year-old mares. She also found that fourth foals were the most successful, and ability gradually declined in subsequent foals.

Ms Barron suggests that, because of these findings, it would seem sensible to mate mares in their prime (if nine years is the peak perhaps we could take 'prime' as being eight, nine and ten years of age) with the best stallions and that we should stop breeding from mares once they reach twelve to fifteen years of age. Of course, we can all think of mares much older than this which consistently produced winners in many spheres. However, we also know that old mares tend to produce lighter, smaller foals than they did when younger, which possibly reduces those foals' performance potential in any discipline.

Although I know of no similar study of the offspring of stallions, this work is so interesting that, if you are breeding for quality rather than quantity, it is worth considering seriously. Lavish the most expensive stud fees on your mare when she is in her prime and stop breeding from her when she is fifteen at the oldest, by which time, as I have said, she may well appreciate returning to work. As for the stallion, try making your mare's expensive mates during her prime time stallions which are also in their prime years, then sit back and wait for your world champion!

Finding a mate

You may have had a personal recommendation of a particular stallion from someone who knows him, you may have seen advertisements for different horses in the equestrian press or via a particular breed society or discipline and you may have spent the summer going round shows and avidly soaking up the stallion classes, particularly the winners, all with the aim of finding a suitable stallion for your mare.

You should have decided just what type of foal you are aiming to breed – which can be anything from a child's pony to a racehorse, a Shetland to a Shire. The type of stallion you put your mare to will obviously make a difference. For example, imagine you have an Arab mare of around 14.3–15hh and you put her to a good Welsh Cob about the same height or slightly bigger; you will get a versatile, middle-weight, cob-like young

teenager's or family mount. If you want to repeat her type, you will naturally put her to a good Arab stallion. If you put her to a Thoroughbred bigger than her, you will get an Anglo-Arab which is quite a different kettle of fish from your cob cross or even your Arab foal – a horse which can compete in Riding or Pony Clubs, do local shows, one-day events, hunter trials, hunting, dressage – you can almost name your discipline. But if you put her to a native pony stallion smaller than she is you will get a quality (but not too fine) child's pony.

So you can see how versatile such a mare is. The same goes for Thoroughbreds and Anglo-Arabs. Warmbloods cross well with Thoroughbreds and Anglo-Arabs, and Irish Draught/Thoroughbred crosses are traditionally superb bases for further Thoroughbred infusions to get competition horses and the first crosses, and also the pure-breds, can make terrific hunters and family riding or driving horses. Native cobs and larger ponies (Welsh Cobs, New Forest and some Highlands) also cross well with hotbloods.

Where many breeders go wrong, as I have said, is to put a nondesdript mare to the nearest, cheapest stallion with no vision of what these two melded together will produce. They so often end up with an equally nondescript horse with no market value which they do not want to keep themselves. So have a clear type in your mind to aim for. Do not just wait and see what comes out.

As for size, there is no problem for the mare in mating her with a smaller stallion but major problems can certainly arise if you choose a horse too big. Most experienced breeders will not put a stallion of two hands or more to a mare, partly because he can seriously and permanently injure her internally during mating and partly because the resulting foal may be so big that it is born prematurely or gives her a very difficult, distressing and even life-threatening foaling. To some extent, the foetus adapts itself to the size of the uterus in which it finds itself, but it is still usually too big for the mare to foal easily, comfortably or safely. I have known several cases of small mares struggling vainly to give birth to over-sized foals, which results in great distress for them and for the foal, and sometimes in the death of one or both. It is safer and more humane to restrict your stallion to one hand bigger than your mare, and to get further height in later generations, if that is what you want.

Make appointments to visit the stallions you have short-listed and take along a good, properly posed snapshot of your mare or maybe a videotape of her to show the stud owner or groom. Although breeding may be somewhat in the doldrums at present, some studs still only accept 'approved' mares partly because they do not want possibly genetically dominant mares producing poor offspring by their beautiful stallion. It is in the interests of stallion and mare owners to match up mates with a view to the utmost compatibility and a clear, honest visual impression of your mare will help greatly in this.

If you cannot keep the appointment, do ring and let them know, not only out of manners because studs are usually very busy places, but also because if you want to try to do business with them again and have let them down, you will have got off on the wrong foot.

It is very important that you feel you can trust the people on the stud and get on with them; you do not want to feel that they are being 'nice' to you solely because they want your money. Most importantly, try to get a good feel for whether or not they are kind people with animals, who will look after your mare and consider her mental needs as well as her physical ones.

The place itself should be tidy and clean, of course, but not necessarily immaculate. The attitudes of the resident horses are important: do they look contented and settled, happy and friendly – or at least not aggressive? Are their boxes or paddocks clean, is the place safe, do they have access to hay or grass and water or are they standing miserably at the backs of their boxes, on thin, dirty beds, in a smelly environment with no food and only dirty water? You do not want that for your mare.

Have a really good look at the stallion and ask if you can handle him before even going to stroke him. Look him over as if you were going to buy him, for in a sense you are. The staff should be proud to show him to you but, because some stallions do bite and it is a stallion characteristic, do not be too worried if they are wary or actually warn you of this. I once owned an Anglo-Arab gelding with the sweetest imaginable temperament by a stallion which was known for biting.

If the stud has more than one stallion at stud and they feel, after seeing your mare's photograph, that a different horse from the one you have chosen would suit her better, let yourself be guided by them. They have a great deal of experience of this sort of thing and are usually well worth listening to.

Once you have had a good look at the stallion and seen him trotted up or, even better, running loose in his paddock, if not actually ridden, and you have had a good look round the place, do not outstay your welcome. Take away any documentation you may want or need and let them know your decision as soon as possible so that you can book your mare to the chosen stallion and set things rolling.

In the past, it was common for mares to 'walk in' to stallions – i.e. they would be led or, more recently, travelled to the horse, mated and brought home again. Then, with the advent of easier horse travel by rail and, this century, by motorised transport, it became the thing for mares to visit more distant stallions and to board at the stud for several weeks to ensure that they foaled safely, were properly in season again and tested in foal before coming home with that year's offspring. When the recession bit in the 1990s, there was another swing back to mares walking in to reasonably local stallions to reduce keep and livery costs, and this still happens a good deal with smaller breeders.

As a novice breeder, it is probably safest, if more expensive, to send your mare to the stud for foaling and covering where she will have expert stud staff and veterinary help on call and be given every chance and facility to produce her foal and become pregnant again. This is an expense well worth while – for the novice breeder, at least.

Some studs provide a video of your mare foaling and a very few will ring you and warn you when they are almost certain she is going to foal, so that you can drive over and be present if you wish, although this is still very unusual.

Chapter 14
Nature
Versus Man

In countries where land is at a premium and horses have been kept in close domesticity for many generations, novice breeders who want to do things the more natural way and let their horses or ponies run and live fairly freely are up against two major stumbling blocks. First, they may simply not have enough land to allow this, and secondly they may be held back by their own entrenched attitudes, or find themselves constantly criticised and warned by other breeders that the natural way is asking for trouble. This could take the form of the stallion sustaining broken bones – usually a leg – from mares kicking him; mares foaling unsupervised and getting into difficulties, mating going on out of control, fighting between mares for the stallion's attention or the stallion killing foals he senses are not his own among other objections.

Some of these arguments may be relevant in some circumstances but can usually be overcome by wise management and common sense – and

there are advantages. Conception rates are almost invariably higher in stock running semi-naturally. Animals, particularly stallions, are generally more content and healthier and therefore often easier to handle and manage. The costs of stabling and intensive labour are reduced because there is less work. Youngstock are also taught natural herd manners and are consequently easier for people to handle and train.

Natural horse society

In natural and feral conditions, there are various horse-devised systems of living, reproducing and surviving. The common picture we all hold in our heads is of a wild and proud stallion, mane and tail flying in the wind on the British moors and hills, the Mongolian tundra or the American prairies. He will have a herd of several mares and their youngsters of varying ages which he

Obviously highly unnatural, regular veterinary examinations are, however, regarded as essential today in the breeding of valuable horses. Examination stocks like these are normally found only on larger studs. Often, the vet has to work over or round the stable door.

A natural herd structure, as in feral life, makes for well-socialised horses and ponies with an acceptance of discipline and respect for older and 'superior' herd members.

protects against all comers, not only other stallions but men and predators, too. He makes all the decisions, gives all the orders and the herd does as it is told. Horses are roamers so we do not think of them as being territorial; they migrate around to wherever the best grass and conditions may be.

This picture is the most popular, and in some respects it is fairly accurate, although studies of equine behaviour and psychology over recent decades, by both professional and competent amateur observers, has shown that behind this picture things are not quite what they seem.

First, we should be clear that there are no truly wild horses or ponies left; there are various feral populations which descend from formerly wild ones (often many generations back) but which have, at some time and in some way, been interfered with by man. Even the herds of rare Przewalski horses reintroduced to their homelands in eastern Asia and 'set free' on large reserves in France and other countries are no longer wild, because they and their immediate ancestors have been living in captivity, albeit a captivity aimed at meddling with them as little as possible with a view to reintroducing them to the wild in the future.

The British and Irish native ponies and cobs now all descend partly from domesticated stock,

the American mustangs descend from escaped or liberated imported horses and so do the Australian brumbies, the New Zealand Kaimanawa horses, the Camargue horses of France, the Polish 'reconstituted' Tarpans and the horses of the Namib Desert in southern Africa (a far from horse-friendly environment). But in fact feral horses, at least those which rarely see people, behave identically to wild ones. So I use the words 'natural' and 'wild' in relation to feral horses and ponies because the word 'feral' can sometimes seem a little pedantic.

It is interesting to note that expert horsemen and observers have argued for generations that once a population of horses or ponies has been domesticated, irreversible changes occur to its mentality and physiology (reduction in brain and gut size have been quoted, decreased 'animal cunning' and intelligence, increased sensitivity to climate and so on) even if no other blood or genes are introduced. However, observers of reintroduced herds of various types have invariably commented that the horses they studied started behaving completely like suspicious, independent wild animals and forming natural herd structures and behaviour patterns within weeks – and sometimes within only days – of being put into the wild, even if previously they had only lived in safari parks or reserves, and had been handled regularly, if not actually petted and trained.

Many owners comment on the fact that their friendly, obedient animals become disconcertingly distant and autonomous when turned out for a holiday of a few weeks provided they have everything they need – ample food and water, enough space, congenial company and shelter, for horses want nothing more, other than to reproduce. They will tolerate visitors and may scrap over titbits (if the humans are unwise enough to take titbits into a herd of loose horses) but they often show every sign of not caring a hoot about their doting 'masters' whom they previously nickered to in welcome in the stable yard.

All this should show us that horses are, underneath their thin veneer of domesticity, still wild animals, still 'ruled by the moon' and still vulnerable to unnatural or stressful practices imposed on them by man.

Old days and nature's ways

Feral and domestic horses and ponies begin to turn their minds (or have them turned by changing hormonal patterns) to thoughts of sex once the natural year has turned after the Winter Solstice, the old Celtic festival of Yule on 21 December, which is the shortest day of the year in the northern hemisphere. Stimulated by lengthening days, the hormones gradually change and tune up the animals' bodies for the changing season and the approach of breeding time.

By February (incidentally another ancient festival, that of Imbolc on 2 February, celebrating the first evidence of the return of life), the horses will certainly be producing their own evidence of spring fever, even though we may regard the season as still midwinter, with the bad weather for which late winter and very early spring are noted. The days are undeniably lengthening, winter aconites and snowdrops are coming through and the horses, if not the humans, are responding by silly behaviour, restlessness and casting their winter coats, and the mares appear to come into season although they will not be fertile yet. Thin from winter's meanness, feral equines are in no physical condition to reproduce yet.

The wheel of the year and the cycle of the hormones gradually work round until, by the third Celtic festival of the year (Oestara, later

Thoroughbred yearlings (and nowadays competition breeds, too) begin work early in life. These Thoroughbred yearlings are being trained to walk obediently in-hand on a stud where their liberty is severely restricted once the time comes to prepare them for the autumn yearling sales. They have to look well-grown, well-groomed and corn-fed to help attract buyers and the best possible prices.

called Easter, on 21 March, the Spring or Vernal Equinox when light and dark are equal) mares will be starting to settle into regular cycles which may still, however, be infertile, depending on the individual. But once the grass begins to show and once mares start actually ovulating, they will accept the stallion's advances. Pregnant mares will obviously not ovulate until shortly after foaling and at this, the foal heat, they may well mate.

The stallion will sense, partly by the mare's flirty behaviour but mainly by the special scents or pheromones she gives off at this time, that she is ready for him. She will approach him, often performing the submissive mouthing, teeth-snapping gesture which indicates that she is his to do with as he will. They graze touching each other, the stallion sniffs and licks the mare's flanks, the root of her tail and her vulva, savouring her scent and flavour with the familiar flehmen gesture. They may mutually groom each other, working with their incisors up and down each other's necks, withers and backs and will smell each other's droppings and urine to confirm identities, to bond and, from the stallion's point of view, to help decide the right moment to mate when the mare is unlikely to break his leg with a kick! The mare will show all the usual signs of being full in season – straddled hind legs, raised tail, winking vulva and small amounts of urine and mucus being passed from her vulva right under the stallion's nose as if to say 'I can't make it any clearer!'

He will usually approach to mate from the side, so as to avoid a kick, enquiring if she really means it, and will then work his way round to her rear, mount and mate. She will co-operate by straddling her hind legs and bracing herself to take his weight. It is all over in a few seconds, when the stallion will dismount and resume close association with the mare until he feels ready to mate her again later.

In feral herds, it has been noted that when two mares are exactly ready for mating at the same time (somewhat unusual in a normal-sized, fairly small herd), the stallion will attend to them both but the two mares may fight over him. The worst time for an outsider mare to try to join a herd is when she is in season, as the resident mares will nearly always be very aggressive towards her at this time – yet this is just when she particularly wants to join and be mated by the stallion. Life can be a bitch, no matter what species you are!

Foaling obviously takes place outdoors and nearly always at night. The mare may drift away from the herd a little to give birth to her foal or may do so near her herdmates. In the latter case, the others, including the stallion and young herd members, often watch the arrival of their new relative with benign interest. It is known, though, that if the foal is by a previous stallion, the present one may sometimes do his best to kill it, and often succeeds, although the longer the mare can protect her foal (in terms of days) the less likely it is to lose its life.

The mare often eats the afterbirth, an action which is believed to be an effort to reduce as far as possible the strong smells of birth, blood, fluids and 'meat' which will attract predators from miles around. Despite the fact that she is not a carnivore, the nourishment it provides will help replenish her energy stores after her exertions.

The foal will be up and suckling, if all is normal, within an hour or so, and by dawn will have mastered its legs sufficiently to enable it to keep up with the dam and the herd reasonably well. Its legs will be almost their mature length and its tiny body will present few problems of balance or weight; it is simply a case of working out which leg goes where, how it straightens and flexes and how to fold them all when it is necessary to lie down and rest.

At first, the mare will be extremely possessive of her foal and may even keep it away from her close friends. Gradually, though, she will allow it a little more leeway and present it to the herd as their new member. It will do its own investigating, the mother permitting, and, over the weeks and months, will spend more and more time away from her, with the other foals, play-fighting and learning all the complex herd manners and social rules essential to life, not to mention absorbing other horses' reactions to other animals and especially to predators. It will take less and less milk until, by the time two more festivals have passed (Beltane, the 1 May celebration of love, and Litha, 21 June, and more familiar today as the Summer Solstice, the longest day of the year which had and has astrological and fateful connotations for many people), the foal, depending on just when it was born, may be getting more nourishment from grass than from its possibly pregnant dam's milk.

Now the days begin to shorten and the horses' brains note this. Although mating and some births will continue, maybe even through to autumn, interest in sex wanes. The old festival of Lughnassad (Lammas) on 1 August is an initial harvest celebration and by the second one, Mabon on 21 September (the Autumnal Equinox when light and dark are again equal), coats will have been casting for a few weeks and the horses will have enjoyed an autumn flush of grass to help them recover from the summer's exertions and to build them up for winter. When Samhain, on 31 October, or All Hallows' Day, on 1 November, announces the start of the hardest season of the year for them, the herd is in non-breeding, energy-conservation mode with lengthening coats, using up body fat and scraping whatever food they can from wherever they can find it. As Yule approaches again, completing the natural year, the mares will be deeply anoestrus, the stallion behaving like an old gelding and the foals taking so little milk that they place no significant strain on their pregnant dams. Natural weaning, which is barely noticeable, takes place a month or two before the mares are due to foal again, although foals and their older brothers and sisters will remain close to their dams in every way, stallion permitting.

At puberty, colts may be banished from the herd by the stallion, which may not want competition. Fillies, too, may be banished, perhaps as a natural way of avoiding close in-breeding, but the observations of researchers and ethologists vary on both these points. Feral herds have been observed in which even sexually mature colts and stallions remain with the family but are denied the right to mate. Sometimes they go off voluntarily and may join a bachelor band of mareless entires, form herds of their own or live a celibate, occasionally solitary life. One report estimated that only one wild entire male equine in a hundred ever mates a mare in his life, which seems a very low figure.

Fillies are popularly believed to stay with their dams and form close bonds for life, but again, reports sometimes state that a stallion will banish his own daughters or at least not mate with them. In-season fillies and mares have been observed leaving herds to find others in which they can mate and reproduce. When their sire loses control of their original herd, they have been known to return to it, along with their own offspring, and renew the bonds with their dam and other relatives.

In this way, it seems, genetic variety is maintained, and close in-breeding avoided. So we should not hold onto a stereotyped picture of the life of a wild herd.

The prices of 'progress'

Life for domesticated populations of horses and ponies is usually quite different from that described above. They are affected in exactly the same way by the lengthening and shortening of day-length but even the results of this (mares coming into and going out of season) can be manipulated by modern veterinary and management techniques so that we can either take our mares' oestrus cycles as they come, bring them on for our breeding plans or stop them occurring to prevent unco-operative behaviour during a mare's competition programme.

Although the horses' instincts will be felt in the same ways, they are often not allowed to respond to them physically as they would in the wild by courting, mating and giving birth according to their own agendas. There are various methods of breeding horses in domesticity, from almost-natural to extremely unnatural, and foals are produced regularly under both systems.

Let us look first at the most unnatural methods, which are almost the norm in the Thoroughbred breeding industry and others in which the animals are worth large amounts of money. These methods may also be found in other studs, where the owners and managers understandably think that because Derby winners and competition world champions are produced by this method, it must be the best. In many such studs, the stallion or stallions are purposely kept away from the mares in a separate stallion box or stallion yard. They may be led out in hand for exercise daily or turned out individually in the securely fenced stallion paddocks but mainly their life is as a stabled, solitary horse, being allowed no close contact with others and, in some cases, not even seeing others except when they are required to serve mares.

The mares have a better existence in that they are normally turned out together a great deal, often living out entirely in summer with their foals. Weaning, however, usually takes place early, at about six months of age, and the foals are kept in same-age groups, usually with no equine adult supervision. When stabled, which may be only at night in winter, they may be alone or in pairs.

From the yearling stage, they are separated also by sex. In the flat-racing sector of the Thoroughbred industry, they will go into training as yearlings. Otherwise, and this applies to

When a mare is full in season she is very likely to stand quite willingly and naturally for the stallion, as this feral mare is doing for her suitor. Domestic mares can be easily accustomed to natural mating with expert management and common sense.

non-Thoroughbreds (competition warmbloods, ponies and others), colts not deemed good enough to be kept entire may be castrated, perhaps as foals but at any rate by three years of age. Breeders may sell their produce at any time from weaning to three years, according to their policies and the prevailing market.

The mating process takes place mainly according to human requirements and beliefs. The stallions and mares may be induced to come into breeding condition by means of exposure to artificial lighting and the mares by hormones administered under veterinary supervision. The horses will be mainly stabled, well fed and may be rugged up, all with a view to persuading their minds and bodies that spring, and breeding, are on the way in the middle of winter, if it is required that they foal early in the following year.

Mares experience a fair number of veterinary examinations to determine when they are likely to ovulate and therefore what the optimal time is for the stallion to serve them. They may be tested (teased) by a gelding or stallion to check whether they accept his attentions and are likely to stand for the stallion to which they are booked. The teaser's reactions will also tell the stud staff whether the mare is really ready for mating.

Teasing is normally done at a specially padded, very strong teasing board or barrier to protect the horses from each other. If the mare is not ready or fully in season, she can react violently to the stallion, lashing out with her hind feet, and can cause serious injury to him and to attendant stud staff. The practice of trying over a stable door is also not uncommon, and reasonably safe, but trying over an ordinary fence or gate is fraught with danger, although it does take place, as horses can easily get their legs through the bars or rails and injure themselves on any non-protective barrier.

Once it is decided that a mare is ready, she is taken to the covering yard or area, probably wearing a bridle or perhaps just a headcollar. She may be twitched, hobbled to prevent her kicking or walking away from the stallion and causing injury to him should he fall or have to dismount quickly, and she will wear padded serving boots on her hind feet to protect him should she kick. She may have a foreleg strapped up and may have a protective bib or shield to protect her withers and neck from the stallion should he bite whilst ejaculating. Her tail will be bandaged to protect his penis from being cut by her tail hairs.

The stallion will be led in, wearing his stallion bridle. He will be held at a little distance from her; the pair will by no means be allowed to court (which would be too dangerous), or even meet. His handler will encourage him to draw (erect his penis) and he will be led up to the mare from behind, usually, or occasionally placed at her hip, will mount and mate and may even be pulled off almost as soon as he has finished.

An attendant (of which there are usually at least three) will throw water (formerly containing disinfectant, a practice now frowned upon) over his penis to cleanse it before he withdraws it, and he will be led away. The mare will be released from her shackles and led around to prevent her straining and passing out the ejaculate, and then returned to her box or field.

As I have said, this is the most extreme of the unnatural domestic methods used to mate horses. There are modifications which can be made to make the process less restrictive. On some studs, the resident stallion does his own teasing, so that he and the mare will at least have met before. Sometimes, he may be led along outside the fences of the mares' paddocks (as nonchalantly as his handler can manage it) or ridden at a walk, being kept a short distance away, and a mental note taken of those mares which approach and show interest in him. These will be tried later.

On some studs, the stallion may be kept in the same yard as the mares booked to him and so mares and stallion get to know each other as they are led about during the day's routine. When stallions are turned out, they may have securely fenced paddocks which allow them to see mares grazing in paddocks a little distance away and very occasionally they may even have their own pony companion mare, which is prevented from coming into season by drug treatment or by having had a hysterectomy.

The mating process, even when carried out in hand, may be made a little less unnatural. If a mare is *fully* in season (and if she is not she is not ready to mate, anyway), she is unlikely to kick meaningfully at the stallion, particularly if she already knows him. Moreover, even if serving boots are still considered necessary, there should really be no need to hobble her or strap up a

foreleg. The bib may still be used to protect her, depending on the stallion's habit. Many very well-run studs with Thoroughbreds and other breeds adopt this less stringent approach perfectly safely.

More and more often, you will see the phrase 'runs with his mares' on a stallion's advertisement or stud card. This means that he and his mares may be out in a herd together, either round the clock or during the daytime, and he mates them naturally when he judges that they are ready. A modification of this may simply be that he is turned out with a mare he knows, or to whom he has already been introduced at least, just for an hour or so to court and mate her. This can happen with his own resident mares, with visiting mares or with those walked or travelled in to him.

There is no breed restriction on this practice. It is understandable that some breeders will not want their mares running with an ill-mannered, even aggressive stallion, but most stallions seem to be made that way out of frustration. I have known of several whose temperaments have totally reformed after being gradually accustomed to more relaxed, natural mating procedures.

Running free is more usual with ponies and cobs than with competition horses, for example, certainly than with Thoroughbreds, although Ireland's National Stud ran a successful experiment with a Thoroughbred stallion in the 1970s in which the horse ran with mares who were sent to him as a last resort to see if this method would get them in foal. I seem to remember that every one of them ended up pregnant.

With this particular horse, it was found that once a mare was ready he would repeatedly mate her, and her alone, ignoring others. It was therefore decided to remove a mare once he had served her, so that his attention would be drawn by others. One person observed the herd's activities all day, noting which mares had been served, then they were removed and newcomers added as they arrived. At night, he and his mares were led in and all stabled individually in the same yard, so being kept as a herd, with the stallion at the head of the yard looking out on his harem. I witnessed this experiment and the horses' management personally, and was impressed at how much more contented they all seemed than horses managed in the more extreme way, despite the fact that there was still a certain amount of interference.

It was particularly interesting to note how willingly the stallion backed off if a mare sent him packing and, conversely, how gradually he enquired of his ladies if they required his services and how gingerly he then approached them. He would come up from the side a few feet away, snort and sniff the mare's muzzle and wait. If she showed the usual signs of acceptance (straddled hind legs, soft facial expression etc.), he would edge round to her hip, ask again, and finally mount and serve her. He did have a few scrapes and scars on his forelegs but no significant injuries, and he very quickly learned, I was told, to keep well out of the way of any mare which showed any sign of unwillingness.

From the mares' point of view, many a shy, reluctant breeder would make approaches to him herself, in which cases he always acquiesced and served her politely. I spent a fascinating afternoon watching them all and talking to the stud observer, and I found the herd's behaviour no different from that of the pony and Welsh Cob herds to which I was used.

Horses, like people and all other animals, have widely differing temperaments. Stallions which have been kept very artificially for many years may well be so psychologically damaged that they will never be safe to 'let go' into a more natural life, but I do think that more attempts could at least be made to see how things go. I do know that many vets, behaviourists and breeders are increasingly favouring natural methods, or at least a more relaxed and less restrictive approach to breeding and mating practices, as many believe that the more unnatural the methods, and therefore the more stressed the horses, the lower the conception rates. A vet and a breeder to whom I spoke some time ago, both of vast experience, both felt that forced and grossly restricted matings (which they believed constituted rape of the mare and which is the normal practice in many studs) produced psychologically impaired foals, and I have heard this opinion expressed frequently since.

However, I cannot advise you which is the best method; you should get as much knowledge and information as you can and then follow your own conscience. There are many very good stallions at stud today in all breeds, types and sporting disciplines, probably too many, so my personal view is that it would be preferable, and quite feasible, to avoid those studs which use the most extreme, unnatural and restrictive practices, both in mating and in general management. If you vote with your feet (which hurts their pockets), who knows, one day they may lighten up and start to be a little more free-handed with their horses and in their techniques without really compromising what they see as safety procedures.

When you approach any stud, make a point of asking if their stallions can run with mares. Whether or not they can, ask if you can come and watch how things are done. Many will not allow this but some may, and you will find it very instructive anyway. Asking around among friends, colleagues or strangers who you know have used that stud will produce a lot of information and help you to make your own decision as to where to send your mare.

Anyone who wants to try running a stallion with mares can get the horses used to the idea by firstly giving the stallion experience at mating in hand, gradually lessening control and restriction and progressing to letting him free with an experienced broodmare for a few hours. Ideally carefully selected, good-natured mares used to running with a stallion should be used. This will teach him how to go on with as little risk as possible. Youngstock raised in a free herd see the stallion mating his mares and accept it as a normal part of life, learning about the 'birds and bees' as readily as their feral cousins.

Studs which have their own resident herds of mares and a stallion may even have their animals, usually ponies, foaling outdoors in herd conditions but this may not be advisable for an introduced, visiting mare, not least because of the possible problem of the stallion, or maybe even the resident mares, trying to kill a new foal. But by all means discuss this with the stud owner if you would like to go this way.

Foaling outdoors is common among heavy, cob and pony breeds and types, but by no means unknown among others, too. The environment (open air, natural ground and so on) may be more appealing provided you are very sure the paddock is safe, and that the weather, according to the type of animal you are breeding, is reasonable. Most people would not wish to foal blood horses of any breed outdoors before late spring or on wet, cold ground.

Other aspects of the 'nature versus domesticity' debate, concerning weaning and management, are dealt with in the next few chapters, but I hope the points raised in this one may at least have set you thinking about the advisability or otherwise of what are often standard practices and about the undoubted benefits of more relaxed and more natural methods.

Chapter 15
Sending Your Mare
to Stud

Sending your mare away to stud, especially if it is her first time and the first time you have been without her, can be almost as nerve-wracking as foaling. There is a good deal of preparation to be done before loading her up and driving her off.

The stud's systems

During your trips to choose stallions, you will have noted the general management of the stud and checked that the place is well run, the staff competent, pleasant and efficient, the paddocks well kept and the stables large, airy and well bedded with clean material. Dust-extracted shavings are a more hygienic bedding than straw which, like hay, always carries some fungal spores, no matter how good the sample. However, straw is normally used on studs as shavings have a habit of getting into the mares' reproductive tracts and of clumping together, not

to mention scattering more easily than straw, and hence not providing such a cushioning footing, particularly in foaling boxes. The latter need really deep, clean beds with thick, high cushioning banks for the protection of foaling mares and newborn foals. They must be thoroughly cleaned out and disinfected between foalings.

Once you have selected your stud and stallion, the nomination form reserving a nomination to (service by) the appropriate stallion, which forms a legal agreement between the stallion owner and the mare owner, must be completed. This must be in order, and the stud and breeder must each hold a copy, before the mare arrives at stud. You may be required to pay a deposit or booking fee beforehand. You must send your completed form to the stud as soon as you have chosen the stallion, remembering that popular horses become booked up very quickly, and you can

Not the safest way to try a mare. Mare or stallion could easily get a leg through the bars of the gate or fence and sustain a nasty injury.

This trying bar is much safer – high, strong and solid. (Vanessa Britton)

only regard your booking as confirmed when you have received back your copy signed by the stud.

If you want your mare to be stabled at night, you will need to give the stud adequate notice so that a box can be reserved and prepared for her arrival. This will be necessary in the case of blood mares foaling fairly early in the year. Others, especially pony, half-bred, cob and heavy mares, will probably be on grass livery during their stay, so check the quality of the herbage and the pasture management during your visit.

It is a good plan to check the feeding arrangements as it is very stressful for mares to have to fight over inadequate food in the paddocks. This is particularly important where the grass is not very good, although every stud's grass should be in good condition. There must be ample safe containers for concentrates (if and when they are fed), hay or haylage, with plenty of space between them. If communal containers are used for feeding concentrates to field-kept horses, try to be present at a feeding time to see how things are run and what arrangements are made to see that each mare gets her fair share. Fighting at feed time can result in seriously injured mares and foals and in general tension and distress, which must be avoided in breeding stock. Ideally, even grass liveries should be caught up and fed individually, as this also gives the staff a better chance to check them over generally and gives the mares a chance to become familiar with the staff.

If the stud will not feed your mare her 'home' brands of feed for reasons of practicality, check what makes they use so that you can accustom her to them well beforehand: the last thing you want is for her to develop colic because of strange feed, or to go off her feed altogether because what is offered to her is unfamiliar.

Also check the facilities for the observation of mares close to foaling and how they are kept. They should be under discreet observation most of the time, especially from dusk to dawn, and not kept with mares which are due to foal at more distant dates as, if a foal arrives early in such company, inquisitive, marauding mares can do it a great deal of damage and cause great distress to foal and dam alike. It has even been known for foals born in such situations to bond with the wrong mare which may or may not have 'stolen' it from its dam, making it extremely difficult to achieve a rebonding with the real dam. Do ask questions about this very important point.

Insurance

The stud will almost certainly insist that you give them written confirmation that the mare is covered by your own insurance policy, and many studs now make owners sign a disclaimer, despite having their own stud insurance. Check with your insurance company whether or not an extra premium is needed to cover your mare and foal during travelling, foaling, covering and whilst they are away from home. Make certain the insurance is all in place before you set off with your precious cargo.

Preparations

In the case of a non-pregnant mare (either a maiden which has never been mated or a barren mare – which simply means she did not become pregnant the previous season, not that she can never have foals), you should have been monitoring her oestrus cycles so that you almost know the precise day when she should be mated. Check with the stud how far in advance of her season they will want her; it is usually a few days to a week before her next season is due to start. Remember that she will stand a better chance of becoming pregnant if she is relaxed and used to her new environment; the staff, the other horses, dogs and so on; the probably different taste of the water, the hay or haylage and grass and maybe the feed, the smell of the place and all the other nuances which only horses could tell us about. She must be relaxed and feel secure in her new 'home' with its different atmosphere and routine. Most mares seem to settle surprisingly quickly when surrounded by other mares in the same situation, on professionally tended, sweet, juicy grazing,

receiving confident, quiet handling and sensing a generally comfortable, competent aura.

Walking in mares obviously gives them no chance to get used to their new surroundings, yet they often still become pregnant. With no statistics on the proportion of pregnancies for walked in or temporarily resident mares, I can only surmise that the latter are probably more likely to conceive than the former, so getting her to the stud a week or so in advance of her season must help.

You will have to book transport or check and prepare your own. Travelling is usually stressful to horses and travelling in-season mares to stud, trying to save a few days' keep, can often make them go off after arrival. You then either have to leave your mare at stud for a further three weeks, by which time she should come into season again, or bring her home for a couple of weeks and take her back – a real nuisance, all for the cost of a few days.

The farrier should be booked to come and remove her hind shoes at least, and dress her feet. If you do not do this the stud will and will add it to your bill. As foaling approaches, pregnant mares will also have their front shoes removed, as no shod horse should be allowed near a foal. Mares sometimes paw their own foals or, if they are clumsy, step on them, and other mares may become aggressive towards strange foals. Shoes will greatly compound any physical injury sustained in this way.

Have all your mare's documentation ready to give the stud staff on arrival – registration documents, vaccination certificates, a careful note of her seasons during the year to date (when she came in and went off and is due to come in again), a note of her worming programme, confirmation from your vet of swabs taken and their results plus any other treatment she has received in the past few months (not essential but sensible).

It will also help the stud if you can give them a brief written history of your mare's breeding performance unless she is a maiden, such as how many foals she has had, whether she foaled easily or had difficulties, any abortions, caesarians or resorbed foetuses, whether or not she is a shy breeder and a good or poor mother, whether she has a good or poor milk supply and so on. Also include details of her diet and any likes or dislikes she has, behavioural quirks and so on.

The stud may require that she arrives with a headcollar bearing her name, particularly if it is a fairly large establishment; they may also attach their own tag containing information for their own use. This is all for identification purposes and saves the staff a lot of time and trouble.

The headcollar will be left on much of the time, at least in the field, and should be a type which will break should it get caught, in other words one of the modern field-safe types which are sturdy but specially made to break under stress, or a lightweight leather one which is also likely to break. The now ubiquitous nylon webbing type with metal fittings are much too dangerous to be left on *any* horse *anywhere*, in my experience, as they will not break when caught up and can result in serious, even fatal, injury.

If the mare has been working during the winter or spring and has been fit, she must have been let down for at least a month before going to stud and preferably longer: very fit mares do not have good conception rates. She should be in very slightly lean to good condition on arrival, certainly neither poor nor fat, and should ideally not have been clipped, so obviating the need for rugs, a real time-waster for busy stud staff to have to deal with. If you must work her during the preceding winter, try to do so lightly to moderately and control her coat growth with extra light and warm rugs (but do not overdo it) so that you do not have to clip her. Otherwise, give her the briefest clip possible and do not clip her after late autumn/early winter. This is not only for her good but also for yours, as the longer her brain takes to imagine it is spring the more difficult it will be for her to conceive.

Similar preparations should be made for mares with foals, but you should not consider travelling a mare with a foal at foot before the foal is at least three weeks old – and then only for a short distance. A proper horsebox should be used (not a trailer) and it should be very well bedded down, with the mare and foal both loose and together in the whole of the travelling compartment, with all partitions and other paraphernalia removed. Avoid having any other animals in the box, unless you have a quiet shared load in a very large equine transporter. You must be able to get from the cab into the horse area without stopping so that you can keep a very close eye on the pair most of the time. Hay-nets are absolutely taboo; hay should be provided on the floor in a corner. Water should be offered regularly during the trip but loose or potentially dangerous containers must not be used.

As with a heavily pregnant mare, the driver must be exceptionally careful, driving relatively slowly, performing all manoeuvres *very* steadily and keeping as much on a straight-line route as possible. Clearly, corners, roundabouts, crossroads and bends in the road will have to be negotiated, but any swaying, bouncing or lurching of the vehicle must be kept to an absolute minimum. Country roads and private drives are notorious for uneven surfaces and potholes so the driver must manoeuvre round these as much as possible to avoid bangs and sudden disturbances. All acceleration and braking must be performed very gradually indeed.

Roundabouts are particularly tricky for the occupants in the back which may first be thrown forwards as the vehicle brakes on approach, then to one side and back as the vehicle accelerates round the island, then to the other side as it reaches its turn-off and finally in the other direction and back as it accelerates on to the straight again. In a nutshell, as one transport expert always says, 'Drive as though you have a full glass of fine wine on the bonnet and you don't want to spill a drop!' Or, in the words of another; 'Drive as though you had no brakes.'

At the stud

Most studs keep maiden and barren mares together in one group, and in-foal mares and those with foals at foot together in another, although the last two are often also separated on a large stud. The stud will study your record of your mare's oestrus cycle and will try her and probably have the vet check her for ripening follicles, at least when the stallion is heavily booked or in the case of valuable Thoroughbred or competition horses, to save the horse's efforts. A veterinary examination is less likely on pony studs, which tend to rely on their stallions' instincts.

Mares are brought up and tried daily on many studs. The mare and the stallion or teaser will be taken to the trying board, with the handlers wearing protective hats, shoes and gloves, and the two allowed to sniff each other over the barrier, nose to nose at first. The mare will be turned sideways on to the board and the stallion allowed to nuzzle her along her back, sniff and nibble her flanks and smell and lick her vulva, just as happens during courting in the wild. If the mare stands for him, is acquiescent, straddles her hind legs and raises her tail etc., it can be assumed that she is well in season and probably

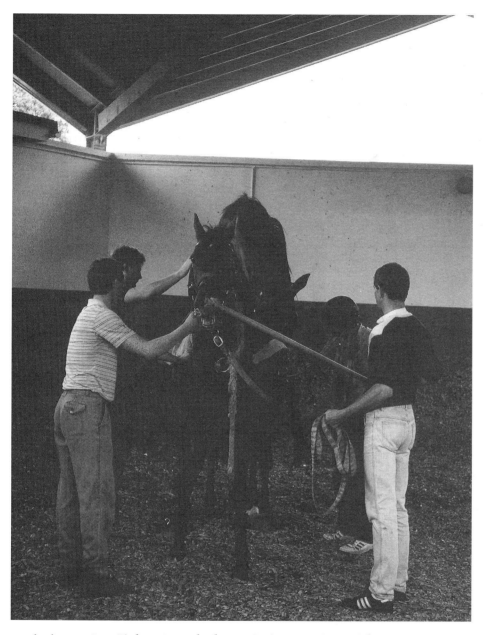

It would be hard to think of a more unnatural and stressful mating procedure than this, yet this is standard practice on most horse, but not pony, studs.

ready for mating. If she rejects the horse, kicking out, maybe biting aggressively at him and obviously not interested, she is probably not ready.

Because ultrasound scanning has fine tuned the information we can acquire on exactly when a mare is likely to ovulate, it is an excellent accompaniment to teasing and, because of its accuracy, many studs now find it necessary to serve a mare only once during each season, according to the vet's interpretation of the scan. But on studs where scanning is not used, mares will probably be served on the third and fifth days of their season.

Shy or nervous mares can be put off by teasing, particularly if the job is rushed or the stallion rather pushy. Such mares often do not show to the stallion and a gentle, tactful horse which takes his time (whether he is her intended or not) can work wonders with such mares, particularly if the humans present adopt a fairly passive attitude. If there is still no reaction, the vet can palpate the mare or scan her to see if a large, soft follicle is present, and advise the stud groom accordingly, who will make the decision to cover her or not, as the case may be.

A description of various covering techniques was given in the last chapter. It is natural for you to want to be present, but few studs will allow this. Some large commercial or national establishments might do so, and may even allow general visitors into the covering yard in a special gallery. Many smaller studs do not have a special covering yard but use a particular area or small paddock. There may also be a barn or covered school or yard which is used for the purpose.

Unless the stallion is running with his mares, he will know what the venue is for. In addition, he will probably always have his same familiar handler and will wear a particular stallion bridle and bit which is worn for no other purpose, and its appearance alone is usually sufficient to get him sexually aroused. Working stallions will be worked in different tack altogether, and will know that mares are not on the agenda when that gear is used. The mare will need no such trigger – being well in season is enough.

Stallion handlers often wear protective headgear, stout shoes and gloves and carry a stick just in case. This does not mean that they are expecting trouble, just that they are prepared for it. Visiting mares can be very unpredictable and, of course, are as strong and potentially dangerous as any stallion, so safety is important. The people who are in charge at covering time must be experienced, confident and able to keep horses under proper control for the safety of all concerned. Training studs obviously have to teach their students, but they will first simply observe several coverings, then perhaps one student at a time will be allowed to help with quiet animals, and progress from there. It is no job for the faint-hearted, the weak-willed or the irresponsible. The watchwords are: calm, firm and positive. The administrative staff will record your mare's coverings on a special certificate which will be needed by your registration authority.

Mares are normally teased again three weeks after covering, meanwhile being watched closely to see if they have come into season. If there is no sign of oestrus and, when tried, the stallion does not show any interest in the mare because she is not giving off her in-season scents, nor she in him, it is likely that she has held to service or 'held at three weeks'. Some owners may feel that this is enough and want to take the mare home to save keep charges. However, there is no guarantee that the mare will stay pregnant at this point, and most studs like to tease again at six weeks. If she still rejects the stallion, it is likely that she is in foal and can go home. Otherwise, the stud will cover her again.

Methods of pregnancy testing were discussed in chapter 5. Whilst at the stud, ultrasound scanning will probably be used to support the teaser's findings or instead of them. Remember that if your mare becomes pregnant the stud has done its job. If she appears to be in foal when scanned at eighteen days or later, you may have to pay the rest of your stud fee, depending on your agreement. If a later scan or rectal palpation subsequently reveals that she is empty (not in foal at that point) – a not uncommon occurrence – the stud may waive the fee or cover her again. Some have a 1 October deadline, with no fee payable if the mare is not in foal on that date even if she has been earlier, which is very generous of them: it is certainly not their fault if a mare resorbs or aborts after all that time. However, this does encourage mare owners to use that stud. Take your vet's, and the stud vet's, advice on pregnancy testing – it is well worth while, no matter what the teaser says!

Your mare should stay at stud for six weeks after her last covering if you can possibly afford to keep her there, as it gives the stud the best possible chance of ensuring that she is in foal. If you must take her away after three weeks (and she has not come into season at her normal time, indicating that she is probably in foal), you may find that any special conditions such as 'no foal, free return' are negated, so check on this.

Once you have returned home, try not to telephone outside any agreed hours or to keep visiting her. If you have chosen the stud wisely there should be no problems and the staff will inform you of any major developments. This does not mean, of course, that they do not want you to ring or visit at all, but do so in moderation.

If your mare is going to foal at the stud (which is advisable, especially if the foal is your first), you need to check when the stud want her to arrive. Travelling very heavily pregnant mares is not a good idea, other than for very short distances, as travelling is stressful at the best of times. Check with the stud and your vet what they advise, but normally it is best to send her four weeks before foaling, not least so that her immune system has time to gear itself up to her new environment, enabling her to pass on the appropriate antibodies to her foal. She will also have this valuable time to get used to the stud and its staff and will hopefully build up a bond of trust with them before foaling.

Chapter 16
Care of the
Pregnant Mare

One of the most noticeable things about a mare when she becomes pregnant is that her temperament often changes. Independent, aloof or 'unpleasant' mares often become more sociable towards humans and other animals and affectionate, interested, 'easy' mares become the reverse, not particularly keen on human attention and perhaps a bit unco-operative. An acquaintance of mine keeps her mare as permanently in foal as possible as it is only in this state that she can handle her! Meanwhile, the mare produces a relentless succession of stroppy, ill-tempered, antisocial youngsters from a variety of stallions. She is obviously genetically dominant as far as temperament goes.

Some mares' appetites also change, and brands of feed they previously enjoyed are rejected, forcing the owner to go through a succession of other makes to find something she will eat. Mares which previously grazed quietly may also now pace up and down the fence, whereas those which were hard to catch and dominated others in the field now come quietly to have their headcollars put on and may even defer to their former underlings.

The first eight months

If your mare has no foal at foot, the first eight months of her pregnancy do not involve any major management changes. It is often advised that newly pregnant mares should be pampered and not stressed in any way, as some say that it is during the first nine weeks or so of pregnancy that resorptions of the foetus or abortions are most likely to occur. A resorption is when the foetal tissues are absorbed into those of the dam so there is nothing to see whereas with an abortion the foetus and associated tissues are expelled through the vagina. If the mare is stabled you will probably find the sad little remains in the bedding but if abortion occurs in the paddock you may find nothing. Furthermore, if the mare does not come into season you may still think that you have an expectant mother

when you do not, as not all mares show a discharge from the vulva after resorption or abortion.

If pregnancy checks show that she is still pregnant, you do not need to worry about any special care during this stage and she can be worked normally provided she is not doing very strenuous work and is not made to sweat. One hears stories about mares whose owners did not even know they were pregnant giving birth to healthy, full-term foals during a season's hard work or – something which was fairly common in the past – a harness mare actually dropping her foal on the whippletree whilst she was working. Needless to say, this sort of thing is not recommended! On the other hand, exercise, as opposed to hard work, is necessary for her own good and for the healthy development of her foal.

Remember that, as time goes on, she will gradually be changing shape and increasing in size. This will eventually start to affect the fit of her saddle and her burgeoning belly will probably push it forwards into the backs of her shoulder blades and also cause the girth to dig in behind the elbows. Although the crupper on the driving harness may counteract this sort of thing in driving mares, it is still something to be watched carefully. Apart from causing her discomfort and possibly pain, and therefore distress, it will interfere with her action. Once it starts happening it is wise to stop riding or driving her altogether. She may also become more and more lethargic, which is another sign that work should stop.

Since anything which distresses her can endanger her pregnancy, it is not a good idea to separate her from her normal friends once she returns home from stud unless one or more of them is likely to seriously harass her. There is certainly no need to separate broodmares from gelding friends just because they are geldings, as is frequently suggested, particularly if they are good friends. Small breeders may have only two horses or ponies and, assuming they are

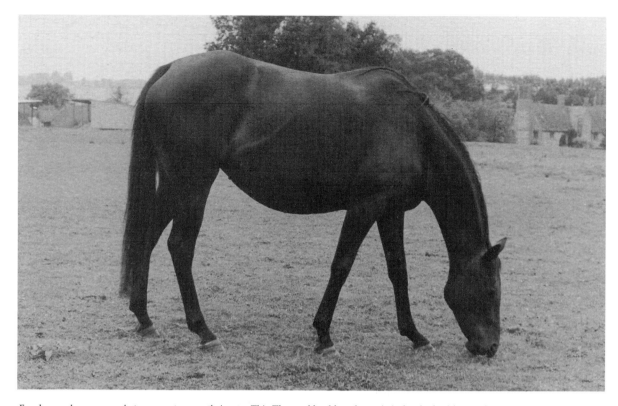

Freedom and grass are what pregnant mares thrive on. This Thoroughbred broodmare is in lovely, healthy condition, neither too fat nor too thin, and is of excellent type and conformation, likely to produce high-class stock from quality stallions of almost any breed.

compatible companions, they should not be separated.

Well-managed grazing with shelter and a good water supply are ideal for pregnant mares, whether they are in work or not. Good shelter facilities enable horses to be out more, saving you considerable time and effort in exercising in-hand or under saddle and in mucking out stables, also saving money on bedding. Animals out a good deal also take perpetual low-level exercise, which is less likely to produce injuries than violent bursts of exuberance when an hour or two in a paddock is their only outlet.

Natural shelter in the form of belts of trees or high, thick hedges are excellent but they do not help against flies, only against driving rain, sleet or snow, or wind and hot sun. There are several long-lasting fly repellants and insecticides available now which will be effective for many hours or even a few days provided the horse does not get wet. Field-safe headcollars can be fitted with fly fringes made from heavy, swinging cotton fronds knotted at their ends, or there is a fringe which does not need a headcollar, just slipping over the ears and round

the throat and coming off easily if caught. Both types are very effective at keeping flies off the face as the horse grazes.

There is always that good old standby the man-made shelter, which is still not made enough use of. My experience is that flies do not frequent shelters much, particularly if you spray the insides with insecticides or use whatever you find keeps them away – cut onions, dried lavender and fly-sticks have all been recommended. Now that there are portable shelters on skids, which can even be moved by the family car when the ground is dry enough, there is no need to deny your animals shelter because you cannot obtain planning permission. Shelters made of straw bales, planks of wood, rope or binder twine and makeshift wooden roofs can be erected with no trouble from planners, and replaced as needed. And, of course, you can always bring in those horses and ponies which are susceptible to extremes of weather.

Over-exposure to foul weather (usually wind and rain), flies – which can create problems by causing horses to gallop relentlessly about (one of the worst things for a broodmare) – lack of shade in hot sun or very muddy ground are all likely to adversely affect your mare's contentment, so normal consideration and good management are even more important now.

But even if your mare lives out mainly, and whether or not she is going to foal at home, you need a decent stable for her, in case of emergencies. As the pregnancy progresses, she will lie down more and you should pay special attention to the cleanliness and thickness of the bed. The floor itself should be slip-resistant as you do not want her, in her bulky, heavy state, to lose her balance or have to fight to keep her feet when getting up from rolling or lying down. Apart from the fact that a fall is very dangerous to mare and foal, any difficulties may well discourage her from lying down at all, so depriving herself of vital rest.

Keeping an under-layer of dirty, impacted bedding is not the way to achieve this, although it is frequently done. By far the best, modern way to ensure a slip-resistant floor is to fit one of the various types of stable matting, usually made of rubber and in different designs. Remember that smooth rubber can itself be slippery when wet; disregard the manufacturers' claims that their flooring will cut your bedding bills, as this is a sales ploy and should not be the main reason one fits such flooring. What you want is a flooring which provides cushioning protection for your mare and foal and is slip-resistant and unmoveable so that the horses cannot disarrange it and trip on it. Look at the various sorts, ask people who have used them and, if possible, go and see them in use before making your choice – other people's standards may not be the same as your own. These floorings should be hosed or swilled down and disinfected just as often as an ordinary floor and, in my experience, almost as much bedding should be put down on them as previously. I find, from observation of other people's horses, that using minimal or no bedding with *any* flooring leads to filthy rugs and horses, sore skin and even skin infections.

The usual general rules apply to stables for pregnant mares: they must be well ventilated without being draughty, light and with more than one outlook for the horse – in other words, as little like a dark, draughty, claustrophobic cell as possible. Size is also important for a broodmare who may well feel the need for more spacious accommodation as she enlarges. The usual size recommended for a foaling box is at least 5m (16ft) square, and a box to accommodate

a horse mare and foal should be very little smaller, although ponies can manage with less.

If your mare lives mainly out, serious grooming should certainly stop, but you must keep a careful check on the condition of her skin to check for parasites or infections. Rolling in the mud will provide plenty of dust in her coat which helps repel parasites and, combined with her natural grease, will protect her a good deal from the elements. Dried-on mud does the same, but this does not mean, of course, that she can be neglected altogether.

Keep a careful watch for mud fever and rain scald developing at any time of year if the weather is wet and remember that pink skin under white hair is the most susceptible, not only on the lower legs. Both these conditions can, if neglected, mean bacteria passing into the bloodstream and can make an animal quite ill, so get on top of them at the first sign. In my experience, these are not first aid situations but should be treated under veterinary advice.

Do not let normal discharges from your mare's eyes or nostrils build up but damp-sponge them away daily, drying the face with an old towel afterwards. Windy weather encourages runny eyes and chapped facial skin, which can become very sore, even infected.

Another area which benefits from regular handling, if not daily sponging, is the mare's udder. Normal skin secretions build up in her 'cleavage', which need removing (I have even seen maggot infestations in mares neglected here). Basic cleanliness is good, but equally important is regular, gentle but persistent handling to ensure that the mare is used to contact and so less likely to object when her foal first tries to suckle. You cannot have no-go areas in things as important as this so habituating her to having her udder and teats regularly handled is important. If she objects (as many maiden mares do), have her held by an assistant in her box, wearing a bridle or maybe a suitable restrainer headcollar and possibly with a foreleg held up, and gently persist for just a few seconds twice a day. Behavioural studies in America have shown that feeding a horse its favourite titbits all the time something unpleasant is taking place, encourages it to associate the action with the reward and ultimately accept it, so get out the mints, carrots or whatever your mare likes for the assistant to give her and gradually train her to accept this treatment. The ideal time is before she is put in foal, of course, in order to minimise distress.

Keep the mane and tail reasonably tidy when grooming, but allow the tail to grow naturally at the top to protect this sensitive area (it will be bandaged anyway for serving and foaling) and allow it to grow long for use as a fly whisk in summer. Shorten it to just below hock length in winter (unless your mare is of a breed which is shown *au naturel*) to prevent its becoming matted with wet mud or snow. You can, of course, shorten natural manes and tails by snapping off the ends of the hairs with your fingertips, and no one will ever guess.

As winter comes on, mares which are out a good deal will grow a thicker, longer winter coat than when stabled, and this is to their benefit. If you bring them in at night, there should be no need to rug them up and they will probably be more comfortable without rugs. It is no kindness to them to encumber them with rugs they do not need – rather the reverse. They can happily withstand very cold temperatures in the dry, still conditions which should be present inside a stable, shed or barn. Only consider rugging very sensitive blood horses and check first that they are actually cold. A hand placed at the base of the ears and on the belly, flanks, loins and quarters, allowing time for body heat to come through the coat to your hand, will usually indicate that the mare is quite cosy inside her natural insulation.

Keep to normal pasture hygiene and follow your vet's advice as to current drugs and worming practices. New drugs occasionally appear which act against different sorts of parasites and require different dosing intervals from other drugs and your vet is the best person to keep you up to date and advise what is best for your situation. The mare's vaccination programme should also be kept up, on your vet's advice, and she should certainly be up to date within a few weeks of foaling so that her foal receives the best protection possible.

Although her environment should be clean and well managed and her food and water of the best quality, there is no need for special diets or feeds at this stage. In fact, it would be a big mistake to overfeed her. She will be healthier if she is in a good to slightly lean condition, although you will start to see changes in her shape as time goes on. If she is not working and is on the right sort of grazing for her type she should not need extra feeding or, conversely, rationing, and will probably be happier and healthier out most of the time, grazing and gently exercising.

It usually takes an experienced eye to tell the difference between a grass belly and a developing foal. At four or five months, the growing foetus moves down lower into the abdomen from the pelvic region, giving the mare a low-slung, enlarged shape which increases with time, whereas a mere grass belly will be more uniform without the low, swinging movement as she walks.

Check that her top line looks and feels well covered but that you can still feel her ribs fairly easily without actually being able to see them. At this stage, if extra feed is needed it would be best to give high-nutrient forages (such as long or chopped alfalfa), haylage depending on the time of year or really good meadow-type hay with a mixture of grasses and, if possible, herbs in it rather than the traditionally favoured seed hay which is more limited nutritionally.

As your mare's belly grows bigger and bigger, she may develop slight hollows in front of her hip bones, which may lead an inexperienced breeder to think she is losing condition. This is quite usual, so be guided by the amount of cover along her spine, back and quarters and, as ever, by how easily you can feel her ribs. If you can actually see her ribs fairly easily, she may indeed be losing condition, so you can take appropriate action. It is better for her to be very slightly lean than overweight, however, although more insulating fat can be allowed in winter.

Her teeth will require the same attention as those of any other horse. She needs to make the best of her grazing and other feed and cannot do so if her teeth or gums are causing discomfort, difficulty or pain.

Even if she is not working, her feet will need very good care when she is pregnant because of the constant and increasing weight she is carrying for most of the year. Laminitis among broodmares is not uncommon, owing to feet being allowed to become too long because of insufficiently frequent trimming, with long toes and low heels. Depending on her particular rate of horn growth, her feet should be trimmed about every four weeks. Any less than this is inviting trouble as the quarters flare out, endangering the structure of the horn tubules and encouraging serious splitting. If the toes grow long and the hoof becomes generally misshapen, great stress is placed on the internal structures and also on the tendons and ligaments of the lower leg, which are already supporting more weight than normal.

If, for any reason, the mare is kept in confined conditions, on damp bedding (usually due to

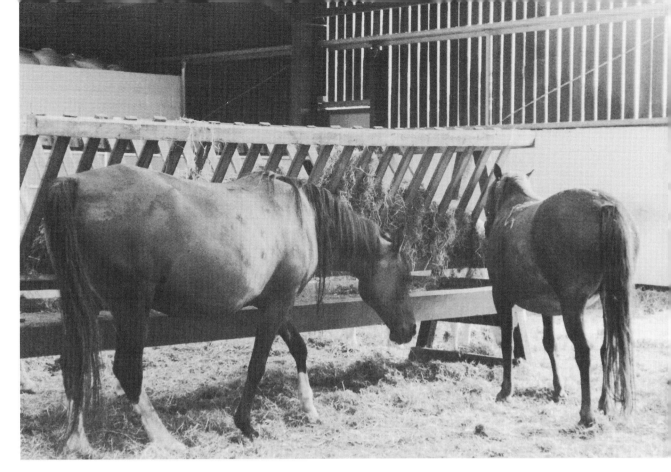

Shelter is important for horses' comfort and peace of mind at any time of year. The doors to this covered yard (out of the picture on the right) are open so that these half-bred hunter broodmares can come and go as they please. It is early spring, so to encourage them to use the yard in inclement weather, hay is provided indoors. However, this type of holder can be dangerous. Sharp corners should be rounded off and gaps filled in to reduce potential injuries.

injurious urine rather than water) or spends much time on wet ground, weakened, softened horn and infections such as thrush are likely. Every effort should be made to keep broodmares (or indeed, any horses) on hygienically maintained, dry bedding and to make sure that those living out in wet weather have somewhere dry and inviting to come in to and stand or lie down in, as broodmares will wish to do more and more as their weight and the strain on their legs and feet increases.

The feet should be thoroughly cleaned out and checked for condition every day. If it becomes necessary, there are a few good hoof dressings on the market now which will help protect the feet without doing long-term harm; ask your farrier and also your vet about these and any new products which appeal to you before making your choice, but do not use ordinary hoof oil on the horn as it can interfere with the natural moisture balance of the feet creating weakened, unhealthy horn.

If the mare does not have a foal with her, you may want to shoe her, at least in front, using quarter as opposed to toe clips to help minimise the spreading of the feet owing to weight-bearing. It should be borne in mind, however, that any broodmare which cannot survive most of the year without shoes (as she will have to do at stud and when her foal is with her) probably does not have good enough feet to be used as a broodmare, as the weakness will quite possibly be passed on to her offspring. Poor horn, flat feet, slow growth and poor natural shape do not bode well for soundness.

Although it is impossible to prove, many people think that unborn foals are susceptible not only to the mare's movements and physical activities, which is reasonable, but also to outside noises and the mare's mental state. Some people believe that the foal is already familiar with the sound of his dam's voice by the time it is born, and maybe even those of her attendants, and that it becomes upset, excited, relaxed and maybe even sleepy or playful when she does. It certainly makes for interesting speculation and, if it is true, it is even more imperative that breeders do everything they can to keep mares well, relaxed and content – and to talk to them!

The last three months

This is the time when your mare should definitely not be working, and should probably be brought in on the winter nights (with a friendly neighbour so she does not spend the night fretting to be with her friends in the field) – unless she is happy to live out with good shelter and is doing well. She should also have her diet gradually changed and the nutrient content increased (see chapter 8), unless she is a native or heavy type maintaining condition as she is.

Keep a very careful watch on her bodyweight, condition scoring her every week and using your instinct and intimate knowledge of her to decide whether or not she is losing or gaining weight. As she develops a bigger belly and changes shape, it can be tricky for a novice breeder to assess her body condition, so get her looked over by a more experienced breeder or, of course, your vet, and compare their opinions with yours.

Particularly towards the end of this period, her quarters may appear to sink a little and the muscles will become softer. She may also start to bag up. Gently rubbing into her udder a light, non-scented oil such as baby oil will help condition the skin and relieve any tightness, as well as continuing your handling training of this important area, but do not overdo it.

She must still have exercise, in-hand if you really cannot arrange anything else for her, partly to help prevent filled legs – which may occur now as a result of an accumulation of fluid – and partly so that she can graze whatever natural herbage may be coming through. Some mares also show a swelling in two rope-like lines (lymph channels) along each side of the belly, which should reduce with exercise.

As the time for foaling draws near, the mare will probably cut down on roughage/fibrous feeds like hay, haylage and forage feeds of her own accord because, as the large intestine where these are mainly digested is in precisely the same part of her body as her developing foal, she will sense that she simply does not have the room inside her for a lot of bulk. She will therefore need more short feeds (concentrates), and the ratio of fibre to concentrates may well now be 1:2 – one-third fibre and two-thirds concentrates.

Her appetite may even fall off a little, so higher-nutrient feeds may be needed to meet the extra demands of the maturing foetus as well as her own. Alfalfa pellets and soaked sugar-beet pulp are good for most animals, particularly those to whom you do not want to feed many cereal concentrates – grains such as oats, barley or maize – and make a good and nutritious mash-type feed. Do not be tempted, however, to feed the old stand-by, bran mashes. These are nutritionally unbalanced and, contrary to former advice, not at all appetising, and we have much better commercial feeds available today.

If your mare is one of those which will not eat alfalfa, even with sugar-beet pulp added, feed her high-energy, high-protein stud cubes with the beet instead, or a specially formulated coarse mix.

Thinly sliced or coarsely grated carrots are great favourites with most animals and are a great appetiser to add to the feeds of a finicky, near-foaling broodmare, and succulents should be provided anyway in winter. Fenugreek (classed as a herbal feed) and linseed are both nutritious and high in energy and condition-giving oils which also help the skin, coat and horn, as does adding a tablespoon of soya or vegetable oil to each feed, gradually introduced as ever.

In nature, of course, food supplies are at their most sparse in the first month or two of this final period, only picking up when the grass starts coming through. The mare can then start making up body condition lost during the winter and, by the time the foal is due, will be in better condition and have grass available to maintain her essential milk supply. This is why, in the wild, most foals are not born until late spring and early summer. In many parts of the UK and Ireland, grass is not worth much until May, yet many breeders want their horses, cobs and ponies born in April to give them a headstart of growth for the showing season.

It seems a shame if breeders are so impatient or, in some cases, so insensitive to their animals' natural biological needs, that the showing of foals cannot be delayed until mid- or late summer to give them the benefit of a later birth and their dams the chance to feed on the very best food of all for equines – good, suitable grass. For some reason we have not yet fathomed (or managed to incorporate into other feeds), there is no bloom quite like a grass bloom on the coat of a well-fed, contented horse. That must surely be worth a few places up the line!

Chapter 17
Foaling
Practicalities

We saw in chapter 6 what you can expect and what normally happens before, during and after the process of parturition or foaling, and in chapter 9 we discussed possible diseases and abnormalities. This chapter is aimed more at you as the breeder, offering a few practical suggestions to help you through a process which you may possibly find rather more stressful than your mare, if you have decided to foal her at home.

If you have an experienced broodmare which has done it all before (which is the best way to start so that you are not both bungling through an unfamiliar process together), she may well sail through everything fairly quickly, showing you what happens and how to proceed. On the other hand, an experienced mare will also be very adept at recognising the signs her body is giving her and will know just when to give birth to ensure that you do not see a thing!

Preparations

If you are wise and cautious, you will have been involving your vet in your mare's pregnancy and will know the dates between which she should foal. Carefully read chapter 6 so that you know the signs to look for and make out a schedule of normal and abnormal events during foaling with a note in the margin of how long each process should take, which you can keep in or near the foaling box. Then, if you are on your own, you will know whether things are progressing normally or whether there is a problem, and can call the vet without delay. Also list all the telephone numbers you may need – certainly the vet's and those of any experienced breeders who may have agreed to help.

Needless to say, if you suspect that foaling is imminent, ring the vet and see whether he or she wants to come out now or wait until the mare actually goes down or the waters break. If the vet does not live too far away, he or she will arrive in plenty of time to take charge of any problems at this stage. If your practice is a good distance away, however, arrange with a nearer practice (via your normal one) for someone else to come out or be available.

If your stabling or field are some distance from a phone, you will have to have a mobile phone. You may also need to arrange for lighting (car headlights or strong lamps) if there is none on site.

Well in advance, prepare a foaling hamper with various essentials and useful items which you may well need. Here is a suggested list, and you will doubtless think of other things:

- your notebook and pen or pencil
- a reliable clock or wristwatch
- a headcollar and leadrope
- clean tail bandages or cohesive bandages kept in wrappers till needed
- soap and towels for the use of the vet and for rubbing down the foal; I assume there is a water supply on site
- hot water in an insulated container in case the vet needs to scrub up
- antiseptic for dressing the umbilical stump
- sterile tape for tying the stump if necessary
- cotton wool
- feed ingredients so that the mare can have a well-earned, mushy reward
- string for tying up the afterbirth
- colostrum substitute or thawed colostrum plus bottle and teat for administering it (prepare for the worst but expect the best)
- a foal rug or old, clean cardigan or jumper in case the foal is sick
- a rug for the mare in case she has had a bad time or goes into shock if there are problems
- a bin-bag or clean, plastic bucket for the afterbirth
- a clean, well-stocked general first-aid kit.

Any specialist items like enemas, a plastic arm glove, lubricants etc. will be brought by the vet.

Some weeks before the foal is due, you will probably move your mare into the foaling box and nearer the time, you should fit it with a low-watt bulb which you can switch on during nightly checks without disturbing her much. If you are expecting a foaling early in the year, you may well find an infra-red heater in the roof a boon, just in case the mare or foal is sick. Needless to say, the box should be thoroughly power-washed or steam-cleaned before you move her in, and not used by any other horse since.

If your mare is living in the foaling box so that she becomes familiar with it and is at ease in possibly new surroundings, you should do a full muck-out and floor-wash every day as she gets closer to foaling, even if you do not normally do so. This includes taking down the high, thick banks you will be providing, sweeping the whole floor and washing down and sprinkling any wet areas with horse-friendly disinfectant. While she is out at exercise in the paddock, let the floor dry and bed it down with clean bedding. The whole point is to maintain hygiene for your vulnerable, new foal. Bacteria and fungi breed very quickly in the relatively humid atmosphere of most stables and foals are much more susceptible to infection than mature horses, so try to avoid it in the first place. The box must be very well ventilated and not draughty.

Foaling boxes must be as physically safe as you can make them, with no floor draughts, no projections, loose buckets or mangers and certainly no hay-nets, which are extremely dangerous where there are young, playful foals. Indeed, there should be no items at all within reach of the foal. Hay should be fed from the ground. The mare's manger and water containers can be at a safe and convenient height for her but must be firmly boarded or bricked off from top to ground so that the foal cannot be injured on them.

Sitting up

Most breeders (at least those without staff) agree that sitting up is the most onerous and wearying part of breeding, even worse than bottle-feeding an orphan foal. On large studs there are proper sitting-up rooms with tea, coffee and snack facilities and often a bed, comfortable chairs and television. On many commercial studs there is closed-circuit television and, where there are several foaling boxes or a full foaling unit, there are banks of monitors or the facility to switch from box to box to check on the mares without disturbing them in any way.

Most novice breeders and small studs, however will have no such facilities, although owners may use their ingenuity to make use of such things as baby alarms and home camcorders linked to portable televisions or computer monitors. There are foaling alarms which work by sensing changes in the temperature and electrical conductivity of the mare's skin. The mare will start to warm up and sweat as the foaling process gets underway, and this is detected by a sensor, usually fitted around the base of the neck and attached to a comfortably fitting roller. The sensor sends a radio signal to a receiver which sounds an alarm within a given radio range – your kitchen, the tack room or anywhere else convenient.

If she is kept away from your home, possibly the best sitting-up arrangements you can make are to use the box next door if you are extremely quiet (although she will certainly know you are there and this may put her off), the tack room, the yard owner's kitchen, or, if there are no outbuildings handy, a camper van, if available, or simply your car. Take a flask or two of whatever you want to drink, sandwiches, chocolate, fruit, a book to read, and wait . . . and wait . . . and wait.

The birth

If you go to check on your mare every half-hour or so, you may find that on one trip she seems to be unconcernedly eating hay and on the next you have an addition to your family. There is a lot to be said for not going in and interfering when all seems to be going well, although this is very difficult for many people. If you *are* around when your foal is born, you will hopefully have some experienced help with you, and will ensure that the amnion has broken, that the foal is breathing and, in due course, that the umbilical cord has broken and you have dressed the stump.

Particularly with a maiden, first-time mare, you can gently drag the foal by its forelegs round to the dam's head so that the bonding process can start and also rub down the foal with the towels or clean straw, although many mares will lick their own foals clean. Watch the mare very closely for signs that she is rejecting her foal, especially a maiden or one which is known to have done this before (not a good breeding prospect) and on no account leave the two alone if you spot signs of it. If in doubt, phone for help and advice.

Try to tie up the afterbirth if you are present, or save it in your bag or bucket so that the vet can check that it is all there and there are no pieces left inside the uterus to set up infection. If you

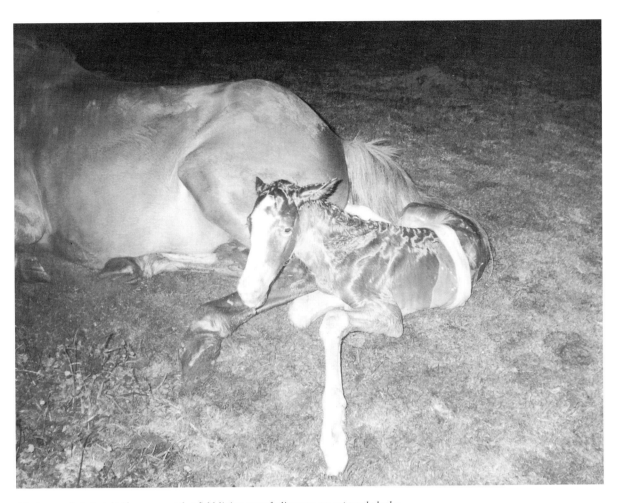

Most mares foal at night if you can catch a field-living mare foaling you are extremely lucky.

are not around, however, and the mare eats it, you will simply have to monitor her apparent health and temperature to keep a check on her instead.

Attendants should stay with or near the mare and foal until both are on their feet and behaving normally, and until the foal has suckled for the first time so that you know it has had some colostrum and, most importantly, that the mare is going to let it feed. This is especially important in a maiden mare or one known to be sensitive about her udder.

Although any restraint of an unwilling mare will upset her, nevertheless she *must* let the foal feed and a bridle and bit, plus a foreleg held up by an assistant, will often do the trick. It is usually best to let the foal find its own way to the udder and a good mother will help it, but if it

seems to be having problems you can guide it slowly with an arm around the buttocks (but never place yourself between the mare and her foal). If the mare is really difficult, you may need help to restrain her, milk her and feed the foal the colostrum from a bottle. You may also need help getting her to accept suckling in future.

If you are having any problems at all, it is best to bring the pair into a field shelter, barn or stable and get expert help at once. Young foals go downhill very quickly if there is trouble so never delay in getting advice and help.

The day after

As most foalings happen at night, once you are sure everything is all right you can go home for some much-needed sleep. If your mare is a

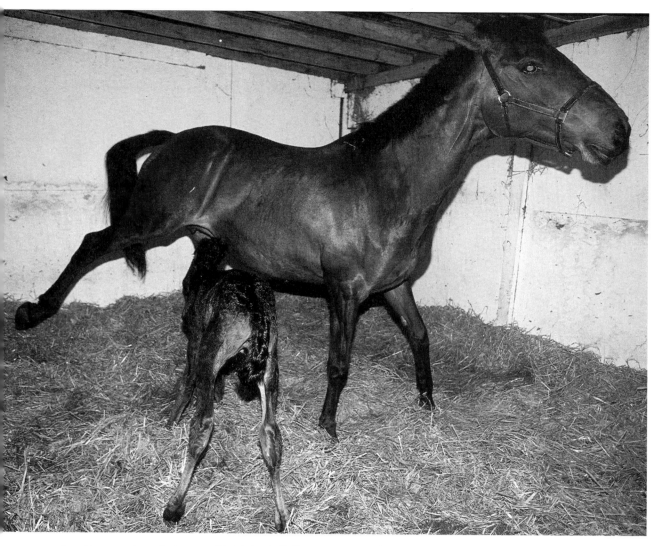

A maiden mare, in particular, may object to her foal's first attempts at suckling, and you may have to step in to restrain her so that she becomes accustomed to it and so that the foal can take his vital colostrum. (Vanessa Britton)

native or heavy type and has foaled in the field, and if the mare and foal are healthy and all is well, by all means leave them in the field for the night. If there are problems or the weather is wet, it would probably be best to bring them in to some kind of shelter. Blood horses foaling in inclement weather should certainly not be out.

In the latter case, watch the weather next morning and, if it is poor, put the pair into a previously arranged alternative turn-out area (an indoor school or covered yard). If it is reasonable, put them outdoors, ideally in a small nursery paddock and definitely away from other mares and foals, for a couple of hours or so, longer if the weather is mild, warm, dry and sunny. The

mare will recover best, and the foal adjust best to the outside world, outdoors in good conditions.

Mother and foal should be checked every three hours or so for the first couple of days, and not just with a cursory glance. Watch their behaviour with each other. If you are worried about anything at all, ring an expert or your vet

As soon as possible, they should be turned out permanently for the spring, summer and autumn but if the weather turns nasty, especially if it becomes wet, and you have no decent shelter, do not hesitate to bring them in, as rain is not at all good for newborn or very young foals. Their coats are far from weatherproof yet and they can easily become chilled or even develop pneumonia.

A few days later

Take some care introducing the new foal and its dam to other mares and foals. They will all foal in their own time, of course, but an ideal arrangement is to have the nursery paddock next to the main one, separated by very secure, safe fencing, so that they can meet without danger of fighting. Then when your new foal is a few days old, try putting it and its dam out with a congenial mare and foal first and progress from there. There is bound to be some squealing, ears laid back, investigation and so on, but most pairs settle in with others whilst still regarding themselves as self-contained units. Over the coming weeks and months, all being well, the foal will make friends and very gradually find its place in the herd.

But what if you do not have a herd? If your mare and foal come home from stud, the mare will be returning to familiar surroundings but the foal will be confronted by a completely new

The correct way to weigh a young foal is for the handler to weigh himself or herself on a pair of bathroom scales first, then to pick up the foal as shown (so as not to injure the breastbone or ribcage) and weigh both. Then, the weight of the handler is subtracted from the total weight of the pair and you have the weight of the foal.

world, strange smells and unfamiliar people and animals and, furthermore, it may suddenly have lost all its new friends. It is always a good idea to try to find another mare and foal as company for yours or even to send yours elsewhere, depending on arrangements, so that the foal can grow and develop normally and have at least one playmate. Many breeders rear sole mares and foals but this is obviously far from natural and it is best for the foal's mental and physical development if it can play normally with at least one other youngster.

BELOW LEFT: Because young foals are 'all legs', they have to adopt all sorts of compromise positions so that they can get their heads down to nibble the grass. (Vanessa Britton)

BELOW: It is a perfectly natural reaction for a foal to run behind its dam when danger threatens. This Welsh Cob filly, though, cannot resist having a peep at the photographer.

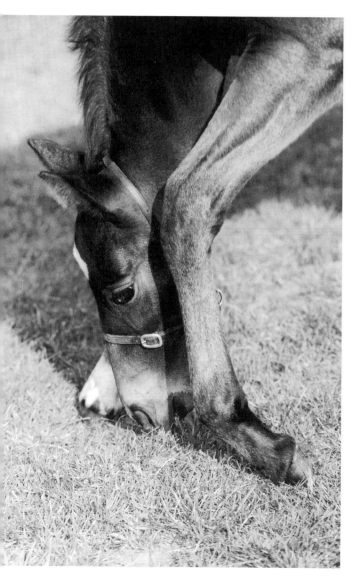

There are bound to be times during your new life as a horse-breeder when you will wonder if you have done the right thing in making yourself responsible for a new life and when you find the commitment in terms of time, money and the possible sacrifice of other aspects of your life a definite disadvantage. Despite the disadvantages, however, most breeders, unless they have had an exceptionally bad time, say that they do not regret a minute of it even if they never breed again.

In the meantime, you have your young foal to consider and plan for, not to mention the dam – although you do not have to make any immediate decisions about what the dam's future entails. If you do not put her in foal again this year, there are always other years, unless her age is against this. If you want to start working

As the foal matures, some people like to gently ride the mare at walk in the field with the foal following along.

Another idea is to ride the mare around the stud with an expert handler leading the foal. It is a good plan to have one extra person available on foot. It should be stressed that most authorities recommend not working the mare at all until the foal has been weaned.

her again, as I have said you can do this very gradually and for very short periods of time once the foal is a few months old, taking less milk and spending more time away from her and playing with friends. It does cause *some* distress to both parties, however, so for the few months it saves, I feel that it is hardly worth it, and work can easily wait until the foal is weaned. Most authorities strongly advise that the mare is not worked until the foal is weaned.

First day

For the first few days, you can expect your mare to be very possessive of her new foal. You may be in a quandary as to whether to leave them alone or whether to use the imprint training system discussed in chapter 6. Contacts are given in the Appendix and it would be a good idea for you to try to talk to as many people as possible

who have used the method and make up your own mind whether or not you feel it appropriate for you – before the foal arrives because you have to start it at birth. If you are going to go the traditional route, however, the first few days will be spent somewhat anxiously and pleasurably keeping a close but discreet watch on your charges – and you can be sure that your mare, whatever your previous relationship, will also be keeping a very close watch on both her foal and you, and on any other humans who try to muscle in. It is not always easy to judge when to insist on 'proper behaviour' and when to leave well alone, so on-the-spot help and advice from an experienced friend or vet will be invaluable to you now.

On their first morning, weather permitting, lead your mare out slowly to your designated nursery paddock or a safe, surfaced exercise area, letting the foal follow her at this stage. If it seems befuddled by the outside world, your assistants (try to have two experienced helpers around) can tactfully guide it with an arm around the rump and breast so that it follows the dam under its own steam. Go into the centre of the area, have the foal very close to the mare's front end and, when they are paying attention to each other, quietly withdraw. Someone should stay within easy observing distance, perhaps cleaning tack on the other side of the fence, while the foaling box is thoroughly mucked out and the floor cleaned and allowed to dry off by leaving all air inlets and outlets wide open (and not letting any other animals at all inside). Clean bedding can then be put over the entire floor, with generous banks and *right up to the back of the door*, and the mare's hay and water can be replenished. Then, after a couple of hours or so, mare and foal can be brought back in.

Handling the foal

After doing the same for a few days, they should be ready to go out with other mares and foals. As the foal finds its feet, co-ordination and strength in these early days, you must start teaching it to lead whilst it is small and young enough to control. Someone should lead the mare slowly while an experienced assistant leads and guides the foal next to her by placing one arm around the rump and holding a stable rubber around his neck with the other hand. No one should get between foal and mare as this could certainly upset them both.

Over a few days, the foal should have a foal slip fitted, and the arm behind the rump should be dispensed with. It can then be led from the rubber around its neck, gradually transferring control to the foal slip. Teach it simple verbal commands all the time, such as 'walk on', 'whoa' and 'stand'. Be consistent with them and with your actions, and you should have the foal leading well under control in a very few weeks. The psychology is to let it think, by using mind over matter and just enough strength to control

Stages in teaching a foal to lead. Initially, at least two people are needed, one to lead the mare and the other to guide the foal with arms round chest and rump, being careful not to get between mare and foal. The next stage is to put a stable rubber or towel around the foal's neck to lead him whilst another helper encourages him along from behind. Eventually, the third person can just be there in case, whilst the foal is led from the rubber and leadrope, progressing to being led from the leadrope only. Finally, it is possible for one person to lead the mare and the foal together.

If foals require their own seperate rations, a special foal creep-feeder can be erected so that the foal can eat in peace without fear of the dam stealing its rations. Food can simply be placed on the ground, like this, or a fixed, rounded and safe container can be used. Creeps on larger establishments can accommodate several foals on similar-type feeds.

it, that you are invincible but benevolent, and you will have sown the priceless seeds of willing co-operation.

The behaviour of mare and foal

The mare's possessiveness of her foal will gradually decrease although, like mothers of any species, mares vary widely in their interest in and influence over their foals. A few continue to be over-possessive and over-controlling whereas others are just the opposite.

When a young foal sees humans, or even other animals, approaching, it is normal for it to hide behind the dam. When it is anxious about anything, it is normal for it to bump her udder with its muzzle to stimulate milk to come down and to have a comforting drink, whether or not it is hungry.

The mare will normally protect her foal against anyone or any animal which bothers it, or even which comes too close whilst it is very young. She will discipline it using her teeth and, as it matures, even her legs and feet if it gets rough

with her or refuses to do as she says.

The foal's investigations of other mares and foals may not be taken kindly, by the mares at least, and they pull no punches, or kicks, where foals not their own are concerned. A foal's dam may intervene if she is close enough, but to a large extent an older foal is on its own, as it must learn what is socially acceptable to other horses and what is not.

Be very careful of the company into which you put your mare and foal. As a small breeder, you may well not have several similar mares and foals (which would be the best company), but foals should really have others to play with for full psychological development. Your mare's former companions (mares or geldings) can certainly stay with her and and her foal, provided they are all friends and unshod. If your mare and foal are being brought up in a natural, all-age herd, maybe even with the stallion present, again simply watch that no one is bullying anyone else. Aggressors should be removed from the main herd for the safety of the others.

Weaning

Many traditionally run studs still wean their foals at around six months, which is increasingly seen by behaviourists, psychologists and some vets and breeders as too young. There are, however, cases when weaning at this stage, or even earlier, may be advisable, for example if a mare is a careless, neglectful mother, if she is aggressive towards her foal or if the foal is unreasonably harrassing and upsetting her as some big colt foals do. Then it is best for both if they are separated. But there is no need to separate them just because a mare has a poor milk supply as the foal's nutritional requirements can easily be made up by means of specialist feeds, not to mention bottle-feeding.

I have already discussed the possibility of lasting psychological damage to early-weaned foals, and more breeders are becoming aware of this and considering more gradual ways of weaning than the traditional one, or leaving dams and foals together longer provided they are getting along. Before six months of age the foal's large intestine is not well adapted to digest fibre, so if the weaning process can be left to about eight months of age it will adjust better and lose less condition.

The usual way of weaning is to decide on the day and, when mares and foals are being brought in from the paddocks, to take the foals to the stables and stable them, either singly or in pairs for company, with the top doors or grilles firmly closed to prevent them jumping out, and to lead the mares away to a distant stable block where they are all out of earshot of each other.

As I say, foals are sometimes put in pairs for

Short-chopped forage feeds are an excellent, nutritious basic feed for almost any horse. They take longer to eat, and involve more chews per kilogram, than traditional, long-fibre feeds such as hay or haylage, and so keep horses occupied for much longer. They mix well with concentrates and soaked sugar beet pulp and, fed in the correct amounts, may be all the supplementary feed needed by ponies, cobs and draught breeds. These feeds can be fed as a partial or complete replacement for hay.

company, but one always turns out to be dominant which makes life even more miserable for the underling. So if this method is used, it is probably best for the foals to be stabled separately but able to see each other. Although they will normally be very distressed for a few days, it is claimed that within a week or so they are not missing their dams at all and soon get over the separation.

They are normally kept in for several days whilst they get over the worst of their unhappiness, then they are turned out together in paddocks well away from the mares. On smaller studs, it may be necessary to send either the mares or the foals to other premises to make sure that they cannot hear each other calling.

The mares are stabled and turned out together as normal, minus their foals and, again within a few days, they normally stop calling for their offspring. They must be dried off (their milk production greatly reduced) by reduced concentrate feeding and by being put on poorer grazing, and they must continue to have exercise to help the dispersal of fluids, swelling and so on. Some recommend lightly milking the mares on a reducing scale for a few days if the udder is very full and tight, but others claim that this continues to stimulate milk production. The udder must be watched for actual tenderness and for the painful infection mastitis. Always consult your vet about any matter on which you are uncertain.

The foal should have been grazing freely by now and, if it is a hot-blood type, it may be eating its own ration of concentrates. Feeding foals from creep feeders (a fenced off area in the stable or paddock under which the foal can walk to get to its own special rations out of reach of the dam) makes sure they get the right sort of feed and can eat in peace without fear of the dam stealing their food – as usually happens! It will start by nibbling his dam's rations but, by weaning, should be eating its own freely. This will help it get over losing its dam.

Feeding roughage and concentrates separately does not mimic in any way the wholesale availability of horses' food in the wild, and it is now recommended that the best way is to mix fibre and concentrates together in one large tub or bowl. This is most easily done by using generous amounts of short-chopped forage feeds

Long-cut, dried alfalfa (lucerne) is a nutritious fibre feed well able to fully replace the best hay and haylage, although quite a few horses find it rather strong in flavour and will not eat it.

When horses are fed in paddocks, it is best for someone to stay and see fair play to ensure that each horse gets its ration and is not bullied. Known bullies (of any age) should be fed out of the paddock or held on a headcollar and leadrope whilst eating, until all have finished. (Vanessa Britton)

Crouching down makes you appear less threatening and more approachable to a foal. (Vanessa Britton)

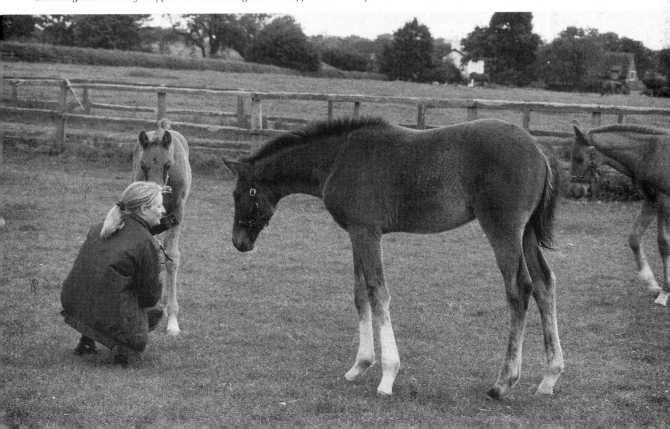

and mixing the concentrate in with it so the horse gets its natural feed with concentrate, which is spread through it, in smaller amounts and over a longer period. This is more natural and effective for horses than specific, separate feeds as we would give ourselves or our dogs. There is nothing to stop an additional supply of hay or haylage being given as well.

The foals often get over weaning better if they are turned out with older youngsters of a friendly temperament. Some breeders find that turning out peer groups (same-age foals) together means that all discipline from elders is suddenly removed and problems with manners and social adjustment can occur.

Youngsters will get over sudden weaning more quickly if they are kept with their normal companions and on their normal routine in familiar surroundings. Small breeders will probably find it best to send the dams away to another stud for a month or so until they have well and truly dried up and can get expert care in the meantime. Leaving mares and foals within earshot is more upsetting, in traditional weaning, than making sure they cannot hear each other calling.

A more gradual way of weaning is to introduce one or two nanny mares into the

It is a good plan for weaned foals and yearlings to be led around to see their home environment and get used to regarding humans as company and herd leaders. Leading from both sides is an excellent plan and helps to prevent the problems which often occur with animals which have been used to being led only from the left, as is traditional but which has no good reasoning behind it.

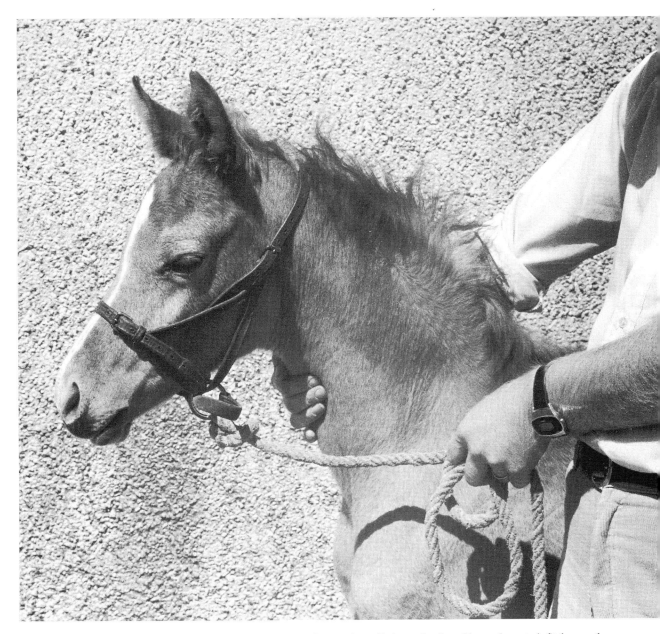

A well-fitting, traditional, leather foal-slip. Foals should be accustomed to wearing a slip from a few days old: some large studs fit them on the first day. Soft, well-maintained leather is kinder to the coat and skin than modern nylon webbing. Cotton webbing is a softer option. Whatever the material, slips must be kept clean and soft, and fitted so that there are no rubs or pressure. It is safer not to leave the slip on all the time, although large studs do so as a time-saver. In that case, the slip must be checked for size almost daily, as foals' heads grow amazingly quickly and slips quickly cause injuries and discomfort if they are too tight.

mares' and foals' paddock so that the foals can get used to them. These will be quiet, friendly mares which do not have their own foals at foot but which are used to youngsters. After a week or so, the dams can gradually be removed at turning out time, over two or three days (the dams of the most forward, independent foals first), or removed altogether, leaving the foals with each other and the nanny or nannies for company and discipline. Where foals are kept in mixed company anyway, the simple exclusion of the dams would constitute a very similar method. Another method, which is preferred by some breeders, is to partition off the loose box

Use your arms and anything you are holding to make yourself seem dauntingly big to a foal trying to evade capture . . .

with the foal on one side and the dam on the other, with their own supplies of food and water, only letting the pair together for occasional suckling.

It is best to discuss different methods and ideas with your vet and other, experienced breeders who have a real feel for horses, but ultimately you will have to decide what best suits your views and facilities.

If foals are eating well and have been accustomed to other company than their dams, either their own older relatives, nannies or friendly geldings, weaning should involve minimal stress, particularly if you can leave it as reasonably long after the traditional six months as you can. If you aim at eight months, the foal will be better adapted physically and mentally to coping without its dam.

Up to three years of age

The continuing handling and training of your foal up to three years of age (after which it may be sold) is very important if you want to be able

150

to offer a well-mannered youngster which will command a higher price. You will find this very much easier if you start gently but firmly disciplining and training it from its earliest days and keeping up the good work.

A foal offered for sale as a yearling or older should lead well in hand, co-operate willingly with the farrier as its feet are trimmed or shod; have good stable manners (moving over and standing to attention as requested), be used to stable and turn-out rugs, if appropriate; be easy to turn out and catch and, depending on circumstances, travel well. It should be used to being groomed and maybe trimmed, must know that biting and kicking are definitely not allowed and that humans are just as definitely for obeying and trusting, certainly not for playing with and trampling on.

If your three-year-old is backed and riding away quite nicely, this is a major advantage for many purchasers, and if you cannot do it yourself it is well worth paying a professional to

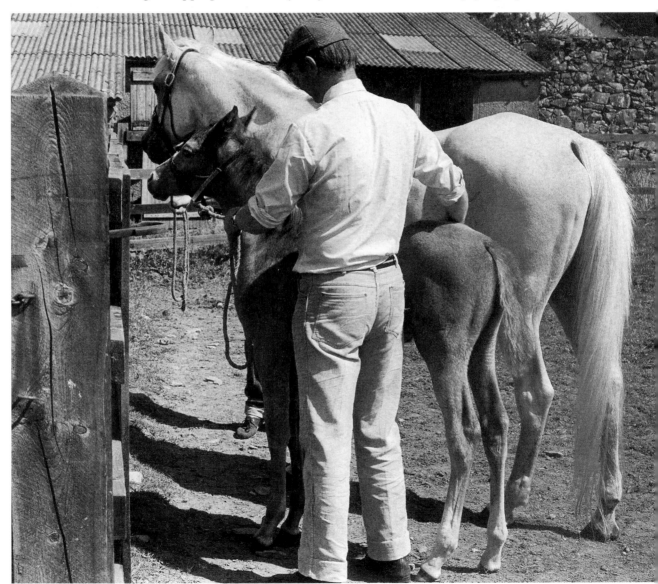

. . . and hem him in with anything available – handlers, the fence and his dam in this case. Should this foal realise that the only way is back, the handler will quickly put an arm behind his tail to block his retreat. The sooner a foal learns that you are a senior member of his herd who always wins (by firm kindness), the sooner he will come to hand as a co-operative, willing young horse who knows where he stands. You will probably be much kinder to him than his herdmates.

151

Never stare back at a foal unwilling to lead and never take a straight pull on the leadrope, like this. The foal's natural reaction will be to shake his head against the rope, pull backwards, possibly rear and fall, with a good chance of his injuring himself.

do it for you, particularly as there is currently a buyer's market which is likely to continue indefinitely. Older youngsters (two and three years) may have worn a roller with side-reins or chambon, depending on their trainer's preferences, and may be lungeing well, with care. Competition horse breeders often loose-jump three-year-olds to check on their abilities but I believe that this should be done sparingly and that no three-year-old should be asked to jump more than 1m (3ft) fences, and only three or four fences about once or twice a week.

Although it is, indeed, a buyer's market, quality stock with good manners will usually find a home. If you have selected your youngster's parents carefully, managed it correctly and well, procured the best veterinary attention and taught it at least basic manners, you have proved to be a responsible, conscientious breeder who will doubtless wish to continue your painstaking work by picking and choosing the foal's future home, whether you sell or lease him, with just as much care and consideration.

Although this foal is now too big to be picked up, we have used her to demonstrate how not to pick up a foal. This method puts too much pressure on the breastbone. Very young foals are picked up with an arm round the chest and buttocks. To weigh them, the handler then stands on a bathroom weighing scale, the weight of the handler is deducted from the total weight and you are left with the weight of the foal.

153

Foals learn about life by copying their dams and other horses. This filly (who, incidentally, has an umbilical hernia which may right itself before treatment is needed) watches with interest as her dam enjoys a good roll. Watch any horse when it rises from rolling to check whether or not it shakes afterwards: if not, this could be a sign of abdominal pain or discomfort. Note the guard fencing round the tree: this prevents horses stripping its bark and killing it. Everything possible should be done to encourage natural shelter and shade from trees and hedges.
(Peter Sweet)

APPENDIX
Useful Addresses

Imprint Training

Imprint Training of the Newborn Foal by Robert M. Miller, DVM, is published by:

The Western Horseman Inc
3850 North Nevada Avenue
Box 7980
Colorado Springs
CO 80933-7980
USA

It is available in the UK by mail order from:

Cloudcraft Books Ltd
16 Wheelers Walk
Paganhill
Stroud
Gloucestershire
GL5 4BW
Tel/fax: 01453 753403

Imprint training of foals is carried out in the UK by:
Ross Simpson
Parelli Natural Horse.Man.Ship
The Natural Animal Centre
Rushers Cross Farm
Tidebrook
Mayfield
East Sussex
TN20 6PX
Tel: 01435 872556; fax: 01435 873268

Other addresses:

The National Foaling Bank
(Miss J.E. Vardon, MBE)
Meretown Stud
Newport
Shropshire
TF10 8BX
Tel: 01952 811234; fax: 01952 811202

British Equine Veterinary Association
5 Finlay Street
London
SW6 6HE
Tel: 0171 610 6080; fax: 0171 610 6823

British Equestrian Trade Association
Wothersome Grange
Bramham
Wetherby
West Yorkshire
LS23 6LY
Tel: 0113 289 2267
Fax: 0113 289 3352

Equine Behaviour Forum
Grove Cottage
Brinkley
Newmarket
Suffolk
CB8 0SF
(Kindly send s.a.e. for information about this voluntary, non-profit-making organisation)

Supplier of high-quality herbal products:

Hilton Herbs Ltd
Downclose Farm
Downclose Lane
North Perrott
Crewekerne
Somerset
TA18 7SH
Tel: 01460 78300
Fax: 01460 78302

INDEX